Thermally C [illegible] 'olumns

Thermally Coupled Distillation Columns

Sustainable and Bio-applications

Juan Gabriel Segovia-Hernandez
Chemical Engineering Department, Universidad de Guanajuato, Guanajuato, Mexico

Eduardo Sanchez-Ramirez
Chemical Engineering Department, Universidad de Guanajuato, Guanajuato, Mexico

Salvador Hernández
Chemical Engineering Department, Universidad de Guanajuato, Guanajuato, Mexico

ELSEVIER

Elsevier
Radarweg 29, PO Box 211, 1000 AE Amsterdam, Netherlands
125 London Wall, London EC2Y 5AS, United Kingdom
50 Hampshire Street, 5th Floor, Cambridge, MA 02139, United States

For accessibility purposes, images in electronic versions of this book are accompanied by alt text descriptions provided by Elsevier. For more information, see https://www.elsevier.com/about/accessibility.

Books and Journals published by Elsevier comply with applicable product safety requirements. For any product safety concerns or queries, please contact our authorised representative, Elsevier B.V., at productsafety@elsevier.com.

Publisher's note: Elsevier takes a neutral position with respect to territorial disputes or jurisdictional claims in its published content, including in maps and institutional affiliations.

Notices
Knowledge and best practice in this field are constantly changing. As new research and experience broaden our understanding, changes in research methods, professional practices, or medical treatment may become necessary.

Practitioners and researchers must always rely on their own experience and knowledge in evaluating and using any information, methods, compounds, or experiments described herein. In using such information or methods they should be mindful of their own safety and the safety of others, including parties for whom they have a professional responsibility.

To the fullest extent of the law, neither the Publisher nor the authors, contributors, or editors, assume any liability for any injury and/or damage to persons or property as a matter of products liability, negligence or otherwise, or from any use or operation of any methods, products, instructions, or ideas contained in the material herein.

ISBN: 978-0-443-33114-5

For information on all Elsevier publications visit our website at https://www.elsevier.com/books-and-journals

Publisher: Priscilla Braglia
Senior Acquisitions Editor: Anita Koch
Editorial Project Manager: Namrata Aggarwal
Production Project Manager: Bharatwaj Varatharajan
Cover Designer: Greg Harris

Typeset by TNQ Tech

Contents

4. Synthesis, design, and optimization strategies in the application of thermal couplings

5. Applications in (bio) thermal coupling processes within a sustainability framework

6. Complete thermal coupling: Petlyuk columns and their implementation as divided wall columns

Acknowledgments

The authors would like to express their deepest gratitude to their mentors, whose guidance, wisdom, and unwavering support have shaped the course of our academic and professional journeys. Their influence has been instrumental in the development of the ideas and concepts presented in this book.

We also extend our sincere thanks to the many students and thesis advisees who have contributed to this field through their hard work, creativity, and enthusiasm. Their questions, experiments, and insights have helped us refine our understanding and push the boundaries of process systems engineering and sustainable distillation.

Finally, we are profoundly grateful to our families, whose patience, encouragement, and unconditional love have been a constant source of strength and inspiration. Their support has sustained us through the many challenges of research, writing, and academic life.

This work is dedicated to all those who believe in the power of knowledge, perseverance, and collaboration.

Preface

This book was conceived in response to the growing demand for sustainable, energy-efficient separation technologies capable of addressing both the environmental and operational challenges of the modern chemical process industry. Distillation remains the most widely used separation technique; however, its energy intensity has prompted significant interest in more efficient configurations, particularly thermally coupled distillation systems (TCDSs).

The evolution of thermally integrated columns—such as side rectifiers, side strippers, Petlyuk, and dividing wall columns (DWCs)—has transformed traditional paradigms in the separation process design. These intensified systems reduce internal remixing, minimize entropy generation, and optimize energy utilization, leading to quantifiable improvements in exergy efficiency, operational costs, and the environmental footprint. Importantly, their relevance is not limited to petrochemical platforms but extends into bio-based industries, circular economy systems, and low-carbon process networks.

Structured across 10 comprehensive chapters, this book provides a multiscale perspective—from thermodynamic fundamentals and sustainability metrics to advanced modeling, optimization, and control strategies. A significant focus is placed on the integration of TCDSs within biorefineries, wastewater treatment, effluent recovery, and renewable fuel purification. Through rigorous analysis and case studies—including 2,3-butanediol, levulinic acid, biobutanol, and biodiesel separations—this book demonstrates how thermally coupled configurations enable energy recovery and valorization in both conventional and emerging chemical processes.

Beyond classical design approaches, this volume incorporates the recent advancements in artificial intelligence, modular design, and process digitalization. It explores neural network-driven decision-making for bioprocess intensification and discusses the implications of electrification, Industry 4.0, and hybrid separation technologies. Furthermore, the final chapters address the practical challenges, opportunities, and future trends in complex DWC schemes and total thermal coupling (TTC), offering a forward-looking view of their industrial scalability and integration with renewable energy systems.

We believe this work will serve as a valuable resource for graduate students, academic researchers, and industrial practitioners interested in sustainable process systems engineering. It offers both foundational knowledge and innovative frameworks to design next-generation separation processes that align with the principles of green chemistry, circular economy, and the 2030 Agenda for Sustainable Development.

Juan Gabriel Segovia-Hernández
Eduardo Sánchez-Ramírez
Salvador Hernández

Chapter 1

Energy as a tipping point in sustainable processes

The historical development of the chemical industry highlights the critical interplay between energy, technology, and societal needs. From its humble beginnings as an artisanal craft to its current role as a global economic powerhouse, the industry has continually adapted to evolving energy paradigms. While its contributions to progress are undeniable, its environmental challenges underscore the importance of innovation and sustainability.

Historical development of the chemical industry

The chemical industry has been a cornerstone of societal and technological progress, transforming agriculture, healthcare, energy, and consumer goods. Its historical evolution is intrinsically linked to advancements in energy sources, material science, and technological innovation. However, while the industry has played a critical role in economic growth, its reliance on energy-intensive processes has posed significant environmental challenges. Understanding its trajectory highlights both the achievements and the hurdles that have shaped its present and future (Dincer & Rosen, 1999).

Early beginnings: Biomass and manual processes

The roots of the chemical industry can be traced back to ancient times, when chemical processes were largely artisanal and based on the exploitation of natural resources. Early practitioners used basic techniques for soap-making, fermentation, dye extraction, and metallurgy. These processes, powered by manual labor and simple combustion of biomass, were inherently limited in scale and efficiency.

For example, natural dyes such as indigo and alizarin were extracted from plants, while fermentation provided rudimentary alcoholic beverages and vinegars. These processes served localized needs, with production scales restricted by the availability of raw materials and the energy required for processing. The absence of mechanized tools meant that the industry's capacity was tightly bound to human and animal labor, with little room for expansion.

Thermally Coupled Distillation Columns. https://doi.org/10.1016/B978-0-443-33114-5.00008-9

The Industrial Revolution: The rise of coal-powered chemical processes

The Industrial Revolution (late 18th to early 19th centuries) marked a significant turning point in the chemical industry. It introduced mechanization and large-scale energy utilization, fundamentally transforming production capabilities. The discovery and utilization of coal as a high-energy-density resource powered this transformation, enabling thermally intensive processes and the creation of entirely new chemicals.

Key innovations during this period included the Leblanc process, developed in the late 18th century, for producing soda ash (sodium carbonate). Soda ash became a crucial input for industries such as glass manufacturing, textiles, and soap production. By the mid-19th century, soda ash production reached hundreds of thousands of tons annually in Europe, a testament to the scalability offered by coal-powered systems. Similarly, the production of sulfuric acid, known as the "king of chemicals" due to its extensive applications, grew exponentially. By 1870, Britain alone was producing over 1 million tons of sulfuric acid annually, primarily for use in fertilizers and industrial processes.

This period also saw the rise of synthetic dye manufacturing. The accidental discovery of mauveine by William Perkin in 1856 marked the birth of synthetic organic chemistry. By replacing natural dyes, synthetic variants like aniline dyes spurred the rapid expansion of textile industries, while also establishing the foundation for modern industrial chemistry.

The 20th century: The petrochemical revolution

The early 20th century introduced petroleum as a transformative energy source, reshaping the chemical industry. Petroleum's versatility extended beyond being a fuel—it also became a feedstock for the production of hydrocarbons, which served as precursors for a wide range of chemicals and materials.

The development of refining technologies enabled the isolation of key compounds like ethylene, propylene, benzene, and toluene. These building blocks fueled the rise of the petrochemical sector, driving the production of plastics, synthetic rubbers, and other polymers. For example, polyethylene and polypropylene, first commercially produced in the mid-20th century, became indispensable materials for packaging, construction, and consumer goods. By 1970, global plastic production exceeded 50 million metric tons annually, reflecting their widespread adoption.

Fertilizer production also underwent a revolution during this period. The Haber-Bosch process, developed in the early 1900s, facilitated the synthesis of ammonia from atmospheric nitrogen and hydrogen derived from natural gas. This innovation supported the large-scale production of nitrogen-based fertilizers, which were instrumental in addressing global food security challenges. By 2000, global ammonia production had reached 100 million metric tons annually, with over 80% used for fertilizers, supporting agricultural productivity for nearly half the global population.

Additionally, electrification enabled the chemical industry to expand its scope. Electrochemical processes powered the production of materials like aluminum, which required significant energy inputs. By mid-century, global aluminum production had reached 2 million metric tons annually, despite requiring around 15 kW h of electricity per kilogram produced.

Environmental costs of growth

The rapid expansion of the chemical industry in the 20th century brought undeniable economic and societal benefits but also revealed significant environmental consequences. Industrial processes became increasingly energy-intensive, with the chemical sector consuming approximately 7% of global energy by the 1980s. Additionally, it contributed to about 18% of industrial CO_2 emissions, primarily from the combustion of fossil fuels for energy and as feedstocks.

Pollution became a pressing issue. Wastewater discharges, toxic byproducts, and air emissions from chemical plants impacted ecosystems and public health. High-profile disasters, such as the 1984 Bhopal gas tragedy and the 1976 Seveso dioxin release, underscored the risks associated with industrial chemical production. Furthermore, resource extraction for feedstocks and the accumulation of persistent pollutants such as plastics and chlorofluorocarbons heightened concerns about sustainability (Gong & You, 2015).

The industry's role in climate change also became evident as CO_2 levels in the atmosphere rose sharply, driven in part by fossil fuel consumption. By the end of the 20th century, the need for more sustainable and energy-efficient practices was apparent, leading to a paradigm shift in industrial operations.

The emergence of green chemistry

The 1990s marked the beginning of a new era, with green chemistry principles providing a framework for addressing the environmental and energy challenges of the chemical industry. This approach emphasized the design of processes and products that minimized waste, reduced toxicity, and improved energy efficiency.

Catalytic processes became a cornerstone of green chemistry, offering ways to lower activation energies and improve reaction selectivity. For instance, zeolite catalysts in petroleum refining significantly reduced energy consumption, saving an estimated 30% of energy compared to traditional methods. Similarly, the development of bio-based processes enabled the production of chemicals like lactic acid, which is used in biodegradable plastics, from renewable biomass.

Life cycle assessment (LCA) tools were adopted to evaluate the environmental impact of chemical processes from raw material extraction to end-of-life. These assessments provided actionable insights for reducing energy footprints and enhancing sustainability.

The 21st century: Renewable energy and circular economy

The 21st century has seen a shift toward renewable energy and circular economy principles in the chemical industry. Renewable energy sources such as solar, wind, and geothermal are being integrated into industrial processes to reduce carbon footprints. For example, solar-powered desalination and wind-driven ammonia synthesis are emerging as viable alternatives to traditional methods.

Carbon capture and utilization (CCU) technologies are also gaining traction. By 2022, global CCU capacity had reached approximately 44 million metric tons per year, with applications ranging from enhanced oil recovery to the production of value-added chemicals. The development of hydrogen as an energy carrier further complements these efforts, with green hydrogen (produced via electrolysis powered by renewable energy) offering a sustainable alternative for processes like ammonia synthesis and methanol production (Hussain et al., 2017).

The chemical industry's adoption of circular economy principles focuses on resource recovery, waste minimization, and recyclability. Advances in chemical recycling technologies are enabling the conversion of plastic waste back into raw materials, addressing the challenges of plastic pollution. For instance, polyethylene terephthalate (PET) recycling rates have reached nearly 60% in regions such as Europe, and the global chemical recycling market is projected to grow to $60 billion by 2030.

Quantifying the industry's impact and potential

Today, the global chemical industry generates over $5 trillion in annual revenue, accounting for more than 15% of global manufacturing GDP. It employs millions of workers and supports countless downstream industries, from agriculture to healthcare. However, its environmental impact remains significant. Efforts to decarbonize and transition to sustainable practices are expected to redefine the industry's trajectory in the coming decades. By 2050, the adoption of renewable energy, green chemistry, and circular economy principles could reduce the sector's CO_2 emissions by up to 90%, positioning it as a leader in the fight against climate change (Hepbasli, 2008).

As the industry navigates the transition to renewable energy and circular economy frameworks, its ability to embrace change will determine its role in building a sustainable future. With energy at the heart of this transformation, the chemical industry remains a vital tipping point for global sustainability efforts.

Energy consumption in the industrial sector and in (bio) separation processes

The industrial sector is one of the largest consumers of energy globally, accounting for approximately 37% of the total energy demand in 2020, according to the International Energy Agency (IEA). Within this sector, the chemical industry alone represents nearly 10% of the total industrial energy use, primarily due to the energy-intensive nature of chemical production and separation processes. Separation operations, including distillation, extraction, and filtration, are particularly noteworthy, as they play a critical role in determining the overall energy efficiency of chemical manufacturing (Moustakas et al., 2020).

The energy demands of conventional separation processes

Separation processes are essential for isolating and purifying products, intermediates, and byproducts in chemical manufacturing. Among these, distillation—a method based on the separation of components by their boiling points—dominates industrial operations. While distillation is highly effective and versatile, its reliance on thermal energy makes it one of the most energy-intensive processes.

Globally, distillation consumes approximately 3% of the world's total energy, and within chemical manufacturing, it accounts for an estimated 40% of the total energy consumption. For example, in the United States alone, energy use for distillation is estimated to exceed 2.5 quadrillion BTUs annually, equivalent to the total energy consumption of some small countries. This substantial energy demand arises from the need to heat large volumes of feedstock to their boiling points and maintain high temperatures during the process (Dincer & Zamfirescu, 2011).

Despite its widespread use, the energy efficiency of distillation is often limited by factors such as:

High reflux ratios: Recycling vapor or liquid within the distillation column increases separation efficiency but adds to energy consumption.
Non-ideal mixtures: Azeotropes and mixtures with closely boiling components require additional energy or specialized techniques for effective separation.

Alternative conventional methods, such as liquid-liquid extraction and adsorption, are less energy-intensive than distillation but have limitations in terms of scalability and selectivity.

Energy profiles of (bio)separation processes

The growing importance of biotechnology and bio-based industries has introduced new challenges and opportunities in separation processes. (Bio) separation methods, which are designed to isolate sensitive biomolecules such as proteins, nucleic acids, and biochemicals, typically have lower thermal energy requirements compared to conventional processes. Key (bio)separation techniques include (Tester et al., 2012).

Membrane filtration

Membrane-based processes, such as ultrafiltration and reverse osmosis, use pressure gradients instead of heat to achieve separation. These methods consume significantly less energy than thermal processes. For example, reverse osmosis for desalination requires approximately 4–6 kW h per cubic meter of water, compared to 10–15 kW h per cubic meter for thermal desalination methods.

Membrane technologies are widely used for concentrating biomolecules, recovering fermentation products, and purifying water in bio-based systems. However, challenges such as fouling, limited chemical resistance, and high maintenance costs remain barriers to broader adoption.

Chromatographic separations

Chromatography is a highly selective method for separating complex mixtures, commonly used in pharmaceutical and biochemical production. Although energy demands are lower compared to distillation, the use of solvents and operational costs can be significant. For instance, large-scale chromatographic separations in monoclonal antibody production require substantial quantities of high-purity water and organic solvents.

Adsorption

Adsorption processes utilize porous materials like activated carbon, zeolites, or metal-organic frameworks to separate components based on affinity. These processes are effective for gas separations, such as CO_2 capture, and liquid-phase purifications. Energy consumption is typically tied to the regeneration of the adsorbent material, which can range from 1 to 3 GJ per ton of CO_2 captured, depending on the regeneration method.

Hybrid and intensified separation processes

The push for decarbonization and energy efficiency is driving the development of hybrid and intensified separation processes. These novel approaches combine multiple separation mechanisms or integrate separation with reaction processes to reduce energy use while enhancing performance.

Hybrid separation systems

Hybrid systems combine traditional methods like distillation with novel technologies to leverage their respective strengths. For example (Omar, 2008):

Membrane-distillation hybrids: These systems pre-concentrate feedstocks using membranes before final separation via distillation, reducing the thermal load.

Pervaporation-distillation: Pervaporation selectively removes components like water from azeotropic mixtures, reducing the energy required for subsequent distillation stages.

Reactive separations

Reactive distillation integrates chemical reaction and product separation in a single column, enabling energy savings of up to 30%–50% compared to separate units. A prominent example is the production of methyl *tert*-butyl ether, where combining reaction and separation reduces capital costs and energy inputs.

Reactive absorption is another approach where CO_2 capture and chemical absorption occur simultaneously, improving efficiency (Olabi, 2014).

Solvent-assisted separations

Solvent-assisted systems, such as liquid-liquid extraction with green solvents, provide energy-efficient alternatives to thermal methods. Ionic liquids and deep eutectic solvents are emerging as environmentally friendly options with lower volatility and tunable selectivity (Kaygusuz, 2007).

Heat integration and process intensification

Heat integration technologies aim to recycle and repurpose waste heat within industrial processes, significantly improving energy efficiency. For example, heat exchangers can recover and reuse up to 50%–70% of energy in distillation systems.

Process intensification strategies, such as dividing-wall columns, further optimize distillation by reducing the number of stages and associated energy demands. Studies indicate energy savings of up to 50% compared to traditional distillation setups.

Scaling challenges and opportunities

Despite their potential, implementing energy-efficient separation technologies at an industrial scale involves several challenges (Kothari et al., 2010):

- Scale-up complexities: Many (bio)separation methods that perform efficiently at the laboratory scale encounter operational challenges when scaled up, such as increased fouling in membranes or pressure drop in adsorption systems.
- Integration into existing infrastructure: Retrofitting hybrid or intensified systems into traditional plants requires significant investment and often faces compatibility issues.
- Economic trade-offs: High capital costs for advanced separation equipment may deter adoption, particularly in resource-constrained regions.

However, these challenges also present opportunities for innovation:

- Advances in materials science, such as the development of fouling-resistant membranes and regenerable adsorbents, are enhancing the robustness and efficiency of separation technologies.
- Digital tools, including process simulation and optimization software, enable more precise energy assessments and system designs.
- Policy incentives, such as carbon pricing and energy efficiency subsidies, are driving investment in sustainable industrial practices.

Quantitative impact of energy-efficient separations

Energy-efficient separation technologies offer substantial potential for reducing global industrial energy use and associated emissions (Segovia-Hernández et al., 2021):

- A transition to hybrid distillation-membrane systems could cut energy consumption in distillation by 15%–25%, resulting in annual energy savings of approximately 1.2 EJ globally.
- Implementing reactive separations across key chemical processes could reduce industrial CO_2 emissions by 250 million tons per year, equivalent to the annual emissions of Spain.
- Expanding the use of low-energy (bio)separations in biopharmaceuticals and biofuels could decrease the carbon footprint of these industries by 20%–40%, depending on the feedstocks and processes involved.
- In monetary terms, these savings translate into billions of dollars in reduced energy costs for industries, strengthening the economic case for adoption.

Future outlook

The ongoing shift toward renewable energy sources and circular economy principles is expected to further transform separation processes. Solar and wind energy can power low-energy separations like membrane filtration and adsorption, decoupling them from fossil fuel reliance. Moreover, the integration of carbon-neutral feedstocks and green solvents aligns with broader sustainability goals.

In parallel, emerging technologies such as machine learning and artificial intelligence are poised to optimize separation process designs. Predictive algorithms can identify energy bottlenecks, simulate process performance, and suggest improvements in real time, accelerating the adoption of innovative methods.

As the industrial sector continues its journey toward decarbonization, the role of separation processes in shaping energy efficiency and sustainability will become increasingly critical. By embracing hybrid systems, reactive separations, and renewable energy integration, the industry can significantly reduce its environmental impact while maintaining economic competitiveness.

In summary, while the industrial sector remains a major consumer of energy, advancements in (bio)separation and intensified processes hold the promise of transforming its energy footprint. The alignment of technological innovation, policy support, and economic incentives will be key to realizing this potential on a global scale (Segovia-Hernández et al., 2021).

Energy as an indicator of sustainability in process design

Energy consumption is a fundamental parameter in evaluating the sustainability of industrial processes. It reflects not only the operational efficiency of a system but also its environmental impact, particularly when nonrenewable energy sources dominate. High energy usage typically correlates with greater greenhouse gas (GHG) emissions and resource depletion, underscoring the need for energy efficiency as a cornerstone of sustainable process design. Beyond direct energy use, sustainability assessments also encompass material efficiency, water usage, waste generation, and overall life cycle impacts, making energy a key driver of systemic improvements.

Energy and sustainability metrics

Energy use serves as a comprehensive metric due to its interconnectedness with other environmental and economic factors (Segovia-Hernández et al., 2023):

- Operational Costs: Energy expenses represent a significant portion of manufacturing costs, ranging from 10% to 50% depending on the sector and process. For example, energy-intensive industries like aluminum production spend 35%–40% of their operating costs on energy. Reducing energy consumption can directly enhance profitability.
- Greenhouse Gas Emissions: Approximately 84% of industrial CO_2 emissions originate from energy use, predominantly from fossil fuels. Energy-

efficient systems reduce fuel requirements, leading to lower emissions and compliance with decarbonization targets.

- Resource Efficiency: Processes with lower energy demands often require fewer raw materials and generate less waste. For instance, high-energy distillation systems not only consume significant energy but also lead to higher wear and tear on equipment, increasing material demand for maintenance.

Frameworks for energy assessment in process design

To integrate energy considerations into sustainable process design, several frameworks, and methodologies are utilized (Segovia-Hernández et al., 2024).

Life cycle assessment (LCA)

LCA evaluates the environmental impacts of a product or process throughout its life cycle, from raw material extraction to disposal. Energy consumption is a central parameter in this analysis, influencing metrics such as carbon footprint and resource depletion.

For example, a comparative LCA of bioethanol production from sugarcane versus corn shows that sugarcane-derived ethanol requires 40% less energy per liter and results in 55% lower GHG emissions, primarily due to differences in feedstock processing and energy recovery potential.

Process intensification

Process intensification involves redesigning systems to achieve higher efficiency and lower energy consumption. Techniques like combining reaction and separation in a single unit (e.g., reactive distillation) can save 30%–50% of energy compared to conventional setups.

An industrial case study on ammonia synthesis demonstrated that incorporating membrane reactors for in-situ hydrogen separation reduced energy use by 15% while increasing ammonia yields by 20%.

Exergy analysis

Exergy, or the measure of useful work potential in a system, offers a nuanced view of energy efficiency by quantifying irreversibilities. Processes with low exergy losses are inherently more sustainable. For instance, replacing conventional boilers with combined heat and power systems improves exergy efficiency from 30%–40% to 70%–80%, significantly reducing fuel requirements.

Designing energy-efficient processes

Sustainable process design leverages a variety of strategies to minimize energy consumption and environmental impact.

Heat integration and recovery

Heat integration optimizes energy use by recovering waste heat and repurposing it within the process. Techniques such as pinch analysis identify opportunities for heat recovery, leading to substantial energy savings.

For instance, integrating heat exchangers in a crude oil refinery can reduce energy demand by 25%–35%, equivalent to saving 10–15 million GJ annually in a medium-sized plant.

Switching to renewable energy sources

Incorporating renewable energy into industrial processes can dramatically reduce carbon footprints. Solar thermal energy, for example, is increasingly used for preheating in cement and glass manufacturing, reducing fossil fuel use by 20%–30%.

A notable example is the use of concentrated solar power in ammonia production, where solar-driven electrolysis generates hydrogen, cutting CO_2 emissions by 90% compared to conventional steam methane reforming.

Electrification of processes

Replacing fossil-fuel-driven processes with electrified alternatives powered by renewable energy is another pathway to sustainability. Electrochemical synthesis methods for chemicals like chlorine and hydrogen peroxide can achieve 50% lower energy intensity compared to traditional methods.

Modular and distributed systems

Modular designs, often implemented in small-scale production systems, allow for localized energy optimization and integration of renewable sources. Distributed biorefineries, for example, use biomass residues from local farms to produce biofuels and electricity, achieving 70%–80% energy self-sufficiency.

Energy-efficient processes and circular economy

The principles of the circular economy, which focus on resource efficiency and waste minimization, are closely tied to energy efficiency in process design.

Material recycling and recovery

Energy-efficient recycling processes, such as chemical recycling of plastics, reduce the need for virgin material production, which is often energy-intensive. For example, producing recycled aluminum requires 95% less energy than producing primary aluminum from bauxite.

Similarly, pyrolysis-based chemical recycling of mixed plastic waste achieves energy savings of 20%–50% compared to incineration or landfilling, while recovering valuable hydrocarbons.

Industrial symbiosis

Industrial symbiosis involves the exchange of energy, materials, and byproducts between neighboring industries to improve overall efficiency. A well-known example is the Kalundborg Eco-Industrial Park in Denmark, where excess steam from a power plant is used by nearby facilities, saving approximately 150,000 tons of coal annually and reducing CO_2 emissions by 240,000 tons.

Valorization of waste streams

Waste valorization converts byproducts and residues into valuable inputs for other processes, reducing both energy demand and waste. For instance, biogas generated from wastewater treatment can replace natural gas as a heat source, achieving net energy savings of 10%–15% in plant operations.

Quantifying the impact of energy efficiency

Energy efficiency improvements in industrial processes yield significant environmental and economic benefits (Segovia-Hernández et al., 2024):

- Global CO_2 Emission Reductions: A 10% increase in energy efficiency across industrial sectors could reduce global CO_2 emissions by 1.3 gigatons annually, equivalent to the combined emissions of Japan and Germany.
- Cost Savings: Energy-efficient technologies in the chemical industry alone could save up to $50 billion annually in operational costs, based on IEA estimates.
- Energy Demand Reduction: The adoption of advanced separation technologies, such as membrane systems, could reduce global industrial energy use by 1–2 EJ per year, a substantial contribution toward meeting net-zero targets.

Challenges in energy-based sustainability metrics

While energy serves as a critical sustainability indicator, several challenges complicate its use (Segovia-Hernández et al., 2021):

- System Boundaries: Defining the scope of energy assessments can vary significantly, impacting results. For example, focusing solely on operational energy use may overlook embedded energy in feedstocks or infrastructure.
- Trade-Offs with Other Metrics: Prioritizing energy efficiency may inadvertently increase water usage or waste generation. For example, some water-saving cooling technologies require more energy-intensive equipment.
- Variability in Renewable Energy Availability: The intermittency of renewable energy sources like solar and wind can complicate their integration into energy-intensive processes.

Future trends in energy and sustainability

Emerging trends and technologies are set to enhance the role of energy as a sustainability metric.

Digitalization and smart systems

Advanced sensors, real-time monitoring, and AI-driven optimization can identify inefficiencies and reduce energy consumption. Digital twins—virtual models of physical processes—are increasingly used to simulate energy-saving interventions before implementation.

Integration of carbon capture and utilization (CCU)

Energy-efficient CCU technologies, such as direct air capture powered by renewables, offer a dual benefit of reducing emissions and supplying feedstocks for value-added products. By 2030, CCU capacity is projected to reach 150 million tons of CO_2 per year, potentially offsetting significant industrial emissions.

Transition to hydrogen economy

Green hydrogen, produced via renewable-powered electrolysis, is poised to replace fossil fuels in high-temperature processes like steelmaking and ammonia synthesis. This transition could reduce industrial CO_2 emissions by up to 20% by 2050.

Energy use is a central metric for assessing and improving the sustainability of industrial processes. From LCAs to innovative process designs, energy efficiency drives reductions in GHG emissions, resource use, and costs. By integrating renewable energy, waste valorization, and advanced technologies, industries can align with global decarbonization goals while maintaining economic competitiveness.

Although challenges remain, the continued focus on energy as an indicator of sustainability will play a pivotal role in shaping a resilient and circular industrial future. The convergence of energy efficiency with broader sustainability principles offers a pathway to a cleaner, more resource-efficient global economy.

Embedded separation processes in the current trends of sustainability and circular economy

The integration of embedded separation processes within the frameworks of sustainability and circular economy represents a transformative evolution in industrial design. By embedding separation operations directly into production workflows or leveraging waste streams as inputs, industries can achieve greater resource efficiency, reduce energy consumption, and minimize environmental impact. These approaches are closely aligned with circular economy principles, where the goal is to close material loops and maintain resources in use for as long as possible (Olabi, 2014).

This paradigm shift is underpinned by advancements in technologies such as in situ separations, electrochemical processes, and renewable energy integration. These innovations are paving the way for sustainable resource recovery, energy-efficient product purification, and climate-resilient industrial practices.

Separation processes and the circular economy

The circular economy aims to replace the traditional "take-make-dispose" linear model with systems that promote recycling, reuse, and resource recovery. Separation processes are integral to this transition as they enable the recovery of valuable materials from waste streams, the purification of intermediate products, and the extraction of reusable components.

Examples of separation processes supporting circular economy principles include:

Metal recovery from waste

Recycling metals such as copper, aluminum, and rare earth elements from electronic waste (e-waste) depends on efficient separation techniques. Hydrometallurgical and electrochemical methods are increasingly used for this purpose.

For instance, the recovery of gold from e-waste using advanced adsorption and leaching techniques achieves recovery rates of up to 95%, significantly reducing the need for primary mining, which consumes 20 times more energy per unit of metal.

Plastic recycling

Chemical recycling methods, such as pyrolysis and depolymerization, use separation processes to convert plastic waste into monomers or other feedstocks. These approaches can achieve energy savings of 30%–50% compared to virgin polymer production.

Recent advances in solvent-based selective dissolution have enabled the recycling of mixed plastics with purity levels exceeding 90%, overcoming a major barrier to high-quality recycling.

Wastewater treatment and reuse

Membrane filtration and advanced oxidation processes play key roles in recovering clean water from industrial wastewater. For example, zero-liquid-discharge systems recover up to 95% of water from effluents, reducing freshwater demand and aligning with circular economy goals.

In situ separation processes: Efficiency through integration

In situ separation processes are embedded directly into production workflows, reducing the need for energy-intensive downstream operations and enhancing resource efficiency. These processes are particularly effective in biotechnological and chemical industries, where product recovery often involves complex and costly steps.

In situ product recovery in bioprocesses

Techniques such as gas stripping, pervaporation, and membrane extraction are used to recover products directly during fermentation or enzymatic reactions. This prevents product inhibition, enhances yields, and reduces energy inputs.

For instance, in bioethanol production, integrating gas stripping for ethanol recovery during fermentation reduces energy use by 20%–30% compared to traditional distillation methods.

Reactive separations

Reactive separations, such as reactive distillation and adsorption-enhanced reactions, combine reaction and separation in a single unit. This integration can reduce the energy footprint by 30%–50% and improve process efficiency.

A notable example is the esterification of acetic acid with ethanol to produce ethyl acetate, where reactive distillation achieves a 40% energy savings compared to conventional setups.

Hybrid bioreactors

Hybrid systems combining microbial bioreactors with in situ separation processes are being developed for bio-based chemical production. These systems, often coupled with renewable energy, enable real-time product recovery and reduced energy costs.

Emerging technologies in embedded separation processes

Several emerging technologies are redefining the possibilities for embedded separations in the context of sustainability and circular economy:

Electrochemical separations

Electrochemical methods use electric fields to drive separations, offering highly selective and energy-efficient alternatives to traditional techniques.

Applications include metal recovery, CO_2 reduction, and water desalination. For example, electrochemical CO_2 reduction systems can produce carbon monoxide or formic acid with energy efficiencies of 50%–60%, serving as precursors for valuable chemicals.

Bioelectrochemical systems

Bioelectrochemical systems combine biological and electrochemical processes for resource recovery and energy generation. Microbial fuel cells, for instance, treat wastewater while recovering bioelectricity and valuable nutrients like ammonia and phosphates. These systems can achieve 50%–70% energy savings compared to conventional wastewater treatment.

Magnetic and acoustic separation

Magnetic nanoparticles and acoustic waves are being developed for highly selective separations in pharmaceuticals and bioprocessing. These techniques eliminate the need for energy-intensive centrifugation or filtration steps.

Advanced membrane technologies

New materials like graphene oxide and ceramic-based membranes are enabling more efficient separations in challenging conditions. For instance, graphene membranes in seawater desalination can reduce energy consumption to 2–3 kW h per cubic meter, compared to 4–6 kW h for traditional reverse osmosis systems.

Renewable energy integration in separation processes

The integration of renewable energy into separation processes enhances their sustainability profile and aligns with global decarbonization goals. Several innovative applications are emerging.

Solar-powered desalination

Solar desalination technologies, such as multieffect distillation and membrane distillation (MD), utilize solar thermal energy to treat saline water. These systems can achieve 50%–70% energy savings compared to fossil-fuel-based methods.

Pilot projects in arid regions, such as Saudi Arabia, have demonstrated capacities of over 10,000 cubic meters per day, providing sustainable solutions to water scarcity.

Wind-driven distillation

Wind energy can power mechanical vapor recompression systems, reducing the energy demand for distillation processes by 20%–40%. This is particularly relevant for remote or offshore industrial operations.

Battery-supported bioseparations

Renewable-powered battery systems enable continuous operation of energy-intensive bioseparations, such as protein purification and monoclonal antibody production. This ensures lower operational costs and reduced carbon footprints.

Quantifying the benefits of embedded separations

Embedded separation processes offer substantial environmental and economic benefits (Segovia-Hernández et al., 2023):

- Energy Savings: Integrating in situ separations can reduce overall process energy use by 20%–50%, depending on the specific application. For example, hybrid reactive distillation systems in chemical plants have achieved annual energy savings of 1.5 TJ per plant.
- Reduced Emissions: By minimizing energy inputs, embedded separations significantly lower CO_2 emissions. For instance, bioelectrochemical systems for wastewater treatment can cut GHG emissions by 30%–40% compared to aerobic systems.

- Economic Gains: Circular economy strategies involving embedded separations, such as metal recovery from e-waste, generate economic value from waste streams. The global e-waste recycling market, valued at $50 billion in 2023, is projected to grow significantly, driven by efficient separation technologies.

Challenges and future outlook

While embedded separation processes offer transformative potential, challenges remain:

- Scale-Up and Integration: Transitioning from laboratory to industrial scales requires overcoming barriers such as fouling, pressure drop, and compatibility with existing infrastructure.
- High Capital Costs: Advanced technologies like bioelectrochemical systems and graphene membranes often involve substantial upfront investments.
- Policy and Regulation: The adoption of innovative separations depends on supportive regulatory frameworks and incentives, particularly for industries transitioning to circular economy models.

Looking forward, advances in materials science, process simulation, and renewable energy integration will drive the adoption of embedded separations. Additionally, global sustainability initiatives, such as the United Nations' Sustainable Development Goals, provide a framework for accelerating the development and implementation of these technologies.

Embedded separation processes are redefining industrial sustainability by aligning production systems with circular economy principles. From in situ product recovery to renewable-powered operations, these methods streamline resource use, minimize waste, and reduce energy footprints. As emerging technologies mature, their adoption will not only address challenges like resource scarcity and climate change but also enhance economic resilience and competitiveness. The continued focus on embedded separations as part of sustainable industrial practices will be instrumental in achieving global sustainability goals and fostering a circular, energy-efficient future.

References

Dincer, I., & Rosen, M. A. (1999). Energy, environment and sustainable development. *Applied Energy, 64*(1–4), 427–440.

Dincer, I., & Zamfirescu, C. (2011). *Sustainable energy systems and applications*. Springer Science & Business Media.

Gong, J., & You, F. (2015). Sustainable design and synthesis of energy systems. *Current Opinion in Chemical Engineering, 10*, 77–86.

Hepbasli, A. (2008). A key review on exergetic analysis and assessment of renewable energy resources for a sustainable future. *Renewable and Sustainable Energy Reviews, 12*(3), 593–661.

Hussain, A., Arif, S. M., & Aslam, M. (2017). Emerging renewable and sustainable energy technologies: State of the art. *Renewable and Sustainable Energy Reviews, 71*, 12–28.

Kaygusuz, K. (2007). Energy for sustainable development: Key issues and challenges. *Energy Sources, Part B: Economics, Planning and Policy, 2*(1), 73–83.

Kothari, R., Tyagi, V. V., & Pathak, A. (2010). Waste-to-energy: A way from renewable energy sources to sustainable development. *Renewable and Sustainable Energy Reviews, 14*(9), 3164–3170.

Moustakas, K., Loizidou, M., Rehan, M., & Nizami, A. S. (2020). A review of recent developments in renewable and sustainable energy systems: Key challenges and future perspective. *Renewable and Sustainable Energy Reviews, 119*, Article 109418.

Olabi, A. G. (2014). 100% sustainable energy. *Energy, 77*, 1–5.

Omer, A. M. (2008). Energy, environment and sustainable development. *Renewable and Sustainable Energy Reviews, 12*(9), 2265–2300.

Segovia-Hernández, J. G., Ramírez-Corona, N., & Aristizábal-Marulanda, V. (2024). *Contributions of chemical engineering to sustainability.* Springer.

Segovia-Hernández, J. G., Sánchez-Ramírez, E., Ramírez-Márquez, C., & Contreras-Zarazúa, G. (2021). *Improvements in bio-based building blocks production through process intensification and sustainability concepts*. Elsevier.

Segovia-Hernández, J. G., Hernández, S., Cossio-Vargas, E., & Sánchez-Ramírez, E. (2023). Challenges and opportunities in process intensification to achieve the UN's 2030 agenda: Goals 6, 7, 9, 12 and 13. *Chemical Engineering and Processing: Process Intensification, 192*, Article 109507.

Tester, J. W., Drake, E. M., Driscoll, M. J., Golay, M. W., & Peters, W. A. (2012). *Sustainable energy: Choosing among options*. MIT press.

Chapter 2

Energy problems in distillation process

This chapter delves into the challenges of high energy consumption in distillation processes, which remain a cornerstone of separation technology in the chemical industry. Distillation is known for its significant energy demands, primarily dictated by thermodynamic limitations inherent to phase change operations. This analysis will cover the thermodynamic underpinnings of these challenges and explore potential solutions through topological modifications of distillation systems.

Conventional distillation columns as a separation alternative

Conventional distillation columns have been the cornerstone of industrial separation processes for over a century, particularly in industries such as petrochemical, pharmaceutical, food processing, and chemical manufacturing. These systems utilize the principle of vapor-liquid equilibrium (VLE) to separate mixtures based on differences in boiling points, with the separation occurring in multiple stages within the column. Distillation columns, despite their widespread application, are notorious for their high energy consumption, a significant concern in the context of increasing demands for sustainability and cost efficiency.

The principle of conventional distillation

The conventional distillation process typically operates under the assumption of ideal vapor–liquid equilibrium, where a mixture of compounds is heated in a reboiler, causing the most volatile component to vaporize and rise up the column. As the vapor ascends the column, it contacts a series of trays or packing, where heat is exchanged, and components condense. The column is designed to maximize the separation between components, achieving a specific product purity at various outlet points. The overhead product is condensed and collected as the distillate, while the heavier, less volatile components are withdrawn at the bottom.

Thermally Coupled Distillation Columns. https://doi.org/10.1016/B978-0-443-33114-5.00013-2

The separation efficiency of a distillation column is governed by factors such as the relative volatility of the components, the number of stages (trays or packing), and the reflux ratio (the ratio of condensed vapor returned to the column to the amount withdrawn as distillate). Despite the well-established design methodology, conventional distillation faces challenges related to energy consumption, particularly because it requires substantial heat input in the reboiler to maintain vaporization and a corresponding cooling load in the condenser (Jobson, 2014).

Energy intensity of conventional distillation

One of the primary drawbacks of conventional distillation columns is their high energy demand. The separation process, particularly for mixtures with low relative volatilities, demands a significant amount of thermal energy. Energy consumption is generally quantified in terms of the reboiler duty, which is the amount of heat required to vaporize the liquid mixture at the column's bottom. The reboiler duty is often the largest single energy expenditure in distillation operations and is highly influenced by factors such as feed composition, column design, and operating conditions.

Studies have shown that in conventional distillation, the energy consumption for separating a mixture of binary components can range from 1.5 to 10 GJ per ton of product. For example, in the case of distilling ethanol from water, which is a common separation in the biofuels industry, the energy requirement can exceed 5 GJ per ton of ethanol produced. This energy intensity is particularly problematic in processes involving heat-sensitive compounds, where excessive temperatures can lead to degradation of the product, further complicating process design and operational challenges (Cui et al., 2024).

Moreover, in many industrial processes, distillation is not a single-step operation but involves multiple distillation columns arranged in series. In such cases, the overall energy requirement can increase significantly due to the additional thermal and mechanical energy input required for each successive stage. A typical multicolumn distillation system for the separation of hydrocarbons, for instance, can demand 20–30 GJ per ton of product, depending on the complexity of the separation and the number of columns involved.

Challenges with energy inefficiency in conventional distillation

The energy inefficiency of conventional distillation systems is driven by several factors. One key issue is the inherent thermodynamic limitations associated with phase change operations. According to the second law of thermodynamics, the separation of components through distillation generates irreversibilities that lead to exergy destruction. Exergy, a measure of the useful work potential of energy, is lost as heat is transferred in and out of the column.

This loss is especially pronounced in the reboiler and condenser, where the heat transfer processes are not perfect.

The energy consumption in distillation systems can be represented by the minimum work of separation, which is the theoretical minimum energy required to achieve a specific separation under ideal conditions. However, real-world systems are subject to inefficiencies, including heat loss, pressure drops, and imperfect heat integration. As a result, conventional distillation often requires energy input that is several times greater than the minimum thermodynamic limit. For instance, in separating ethanol from water, while the minimum reboiler duty may be 1.5 GJ per ton, typical industrial systems consume around 5–6 GJ per ton of ethanol produced, reflecting a considerable efficiency gap (Halvorsen & Skogestad, 2011).

Additionally, distillation columns are typically designed for optimal performance under a specific set of operating conditions. Variability in feed composition, flow rates, and temperature can result in suboptimal operation, leading to higher energy consumption. In cases where feed compositions change frequently or include impurities, the column must adjust to maintain separation performance, often increasing energy usage and reducing overall process efficiency.

Alternatives to conventional distillation

Given the energy-intensive nature of conventional distillation, significant efforts have been made to explore alternative separation technologies that can reduce energy consumption and improve the overall efficiency of the separation process. Some of the most promising alternatives include membrane technologies, adsorption processes, and pressure swing distillation (Janković et al., 2024).

Membrane technologies

Membranes are being explored as an alternative to conventional distillation for a variety of separations, particularly in cases where the difference in volatility between components is small. Membrane separation processes, such as pervaporation and nanofiltration, allow for selective permeation of components through a semipermeable membrane, separating them without the need for phase changes. While membrane technologies have shown promise in reducing energy consumption, they are still limited by factors such as membrane fouling, selectivity, and scalability for large-scale operations.

Adsorption processes

Adsorption technologies involve the selective binding of components to a solid adsorbent material. These processes have been explored for applications such as natural gas purification and water desalination. Adsorption processes can be more energy-efficient than distillation because they do not require phase changes; however, they are often limited by the need for regeneration cycles and the capacity of adsorbents.

Pressure swing distillation

This is a variation of conventional distillation that takes advantage of the fact that the relative volatility of components in a mixture can change with pressure. By operating at different pressures in different stages of the column, pressure swing distillation can sometimes achieve the same separation with lower energy input, particularly for mixtures where the relative volatility changes significantly with pressure.

Despite these alternatives, distillation remains the dominant separation technology in many applications due to its well-established operational methodology and versatility. Even in cases where alternative technologies are being explored, distillation is often employed as part of a hybrid system, where its ability to handle large volumes and produce high-purity products complements the strengths of other technologies.

Energy recovery and process intensification in distillation

To address the energy inefficiencies inherent in conventional distillation, several strategies have been developed to optimize energy usage and improve the overall sustainability of the process. These strategies fall under the umbrella of process intensification, which aims to enhance the performance of industrial processes while minimizing energy and material inputs (Kim, 2016).

Heat integration

One of the most common approaches to reduce energy consumption in distillation is through heat integration. This involves the recovery of waste heat from the hot stream exiting the distillation column and using it to preheat the cold feed entering the column or to supply heat to other parts of the process. Techniques such as heat exchangers and the use of external thermal energy storage systems are widely implemented to improve the overall heat economy of the system. In some cases, heat integration can reduce energy consumption by 10%–30% in large-scale distillation operations (Kiss et al., 2012).

Divided wall columns (DWC)

A divided wall column (DWC) is a more advanced column design that integrates multiple distillation stages into a single column, allowing for improved separation efficiency and reduced energy consumption. The DWC can simultaneously perform the functions of multiple conventional distillation columns by utilizing internal walls to separate different streams. This design has been shown to reduce the reboiler duty by up to 40% compared to traditional distillation columns in certain applications (Skiborowski, 2023).

Reactive distillation

In reactive distillation, chemical reactions are integrated into the distillation column. By combining reaction and separation in a single unit, it is possible to reduce the number of stages required for separation, resulting in a lower reboiler duty. This process has found applications in the production of biodiesel and the synthesis of certain chemicals, where the reaction itself drives the separation of products. Reactive distillation can reduce energy consumption by as much as 20%–50% in certain processes applications (Skiborowski, 2023).

Despite the high energy consumption associated with conventional distillation columns, they remain a critical tool in industrial separation due to their robustness, scalability, and versatility. While alternative technologies such as membrane processes, adsorption, and pressure swing distillation offer potential energy savings, distillation continues to dominate many separation applications. To mitigate the energy inefficiencies inherent in distillation, innovations in column design, heat integration, and process intensification offer promising avenues for improving performance. As industries continue to face increasing pressure to reduce energy consumption and improve sustainability, further advancements in distillation technology will be essential to meet these challenges.

Thermodynamic efficiency as a performance indicator of a distillation column

Thermodynamic efficiency is a fundamental metric for assessing the performance of a distillation column, which is a widely used separation technology in chemical and petrochemical industries. Distillation involves the separation of components in a mixture based on differences in their volatilities. The column is typically designed to separate liquid mixtures into their individual components by utilizing heat to vaporize the more volatile components and condense them back into liquid phases at various levels of the column. However, the efficiency of this process is not solely determined by the purity of the separated components but also by the energy required to achieve that separation (Stichlmair et al., 2021).

Definition and importance of thermodynamic efficiency

Thermodynamic efficiency can be defined as the ratio of the minimum work (or minimum energy) required for a separation process to the actual energy consumed in the distillation column. This ratio serves as a performance indicator, allowing engineers to evaluate how well a distillation column performs relative to the theoretical minimum energy requirements.

In distillation processes, the minimum work refers to the least amount of energy necessary to achieve a separation between components, typically derived from the concept of reversible processes in thermodynamics. The actual energy input, on the other hand, includes all forms of energy consumed, including heat and mechanical work, that are required for the operation of the column. The larger the discrepancy between the actual energy input and the minimum required energy, the less thermodynamically efficient the process is, signaling inefficiencies in the system (White, 2012).

The Carnot limit: An idealized benchmark

The Carnot efficiency serves as an idealized upper bound on thermodynamic efficiency in many thermal processes, including distillation. The Carnot efficiency is defined as:

$$\eta\text{Carnot} = 1 - \text{Th}/\text{Tc}$$

where:

- ηCarnot is the Carnot efficiency,
- Tc is the absolute temperature of the cold reservoir (in Kelvin),
- Th is the absolute temperature of the hot reservoir (in Kelvin).

For a distillation column, the temperatures of the vapor and liquid phases in the column act as the hot and cold reservoirs, respectively. The Carnot efficiency defines the theoretical maximum efficiency of the heat exchange process within the distillation column. Achieving this maximum efficiency would imply that no exergy is lost during the transfer of heat between the different stages of the distillation column, which is an idealized scenario. However, due to the inherent irreversibilities in the distillation process, such as friction, nonideal heat transfer, and mixing of vapor and liquid phases, the actual performance is always below the Carnot limit (Altay et al., 2025).

Exergy destruction and irreversibilities in distillation

Exergy destruction is a key concept in understanding the thermodynamic inefficiencies of a distillation column. Exergy refers to the maximum useful work that can be obtained from a system as it comes into equilibrium with its surroundings. In real processes, exergy is destroyed due to irreversibilities, such as friction, heat losses, and imperfect mixing of phases. The destruction of exergy in a distillation column is a primary cause of inefficiency.

Irreversibilities in distillation arise from several sources (Altay et al., 2025):

- Heat transfer inefficiencies: In practical systems, heat exchangers often operate below their ideal performance due to temperature gradients, material limitations, and fouling, leading to significant exergy destruction.
- Mass transfer inefficiencies: The mass transfer rate between vapor and liquid phases can be suboptimal, which reduces the efficiency of the separation process.
- Pressure drop: The pressure drop across the column can lead to additional energy consumption for pumping fluids, contributing to inefficiency.
- Vapor and liquid mixing: In an ideal column, vapor and liquid phases should be perfectly separated at each stage. However, due to turbulence, diffusion, and other factors, mixing occurs, which reduces the effectiveness of the separation.

The total exergy destruction in a distillation column can be quantified by examining the exergy balances for the system, considering the incoming and outgoing streams of heat and mass. The exergy destruction rate is directly related to the difference between the ideal and actual performance of the column, which is linked to the thermodynamic efficiency.

Thermodynamic efficiency in industrial distillation columns

In industrial applications, distillation columns are designed to achieve specific separations, such as purifying chemicals, refining petroleum, or producing high-quality solvents. However, the actual thermodynamic efficiency of these columns often deviates from the theoretical ideal.

A study by Kiss et al. (2012) found that the thermodynamic efficiency of a typical distillation column operating in the chemical industry is often less than 10%. This is primarily due to the significant energy losses associated with the heat integration, pressure drop, and inefficiencies in the mass transfer. Another study by Kiss and Ignat (2013) found that for a distillation column used in the production of ethanol, the thermodynamic efficiency was approximately 8%, with most of the energy lost due to high reboiler temperatures and poor heat integration strategies.

The use of advanced energy-saving techniques, such as heat pumps, advanced control systems, and energy-efficient heat exchangers, has been shown to improve the thermodynamic efficiency of distillation columns. For example, the integration of a heat pump can reduce the reboiler duty, leading to lower energy consumption and higher thermodynamic efficiency. In some cases, efficiencies as high as 30%–40% have been reported, though this is still far from the ideal Carnot efficiency.

Optimization of thermodynamic efficiency in distillation

To improve the thermodynamic efficiency of distillation columns, several optimization strategies can be applied. These strategies are focused on reducing exergy destruction and minimizing energy consumption (Yang et al., 2020).

- Column design optimization: One of the most effective ways to improve thermodynamic efficiency is by optimizing the column design. This includes adjusting the number of stages, the column diameter, the type of packing or trays used, and the vapor–liquid contact area. Proper design can reduce pressure drops, minimize heat losses, and increase the efficiency of mass transfer.
- Heat integration and recovery: Heat integration techniques, such as the use of heat exchangers or heat pumps, can reduce the overall energy demand of the distillation process. By recovering waste heat from the system and reusing it for preheating the incoming feeds, the thermodynamic efficiency can be significantly improved.
- Operating conditions optimization: The operating conditions of the column, such as the reflux ratio and reboiler temperature, can be adjusted to minimize energy consumption. However, this must be done without compromising the purity and yield of the separation. Lowering the reflux ratio, for instance, reduces energy consumption but may affect the separation quality, requiring a careful balance.
- Advanced control systems: The implementation of advanced control strategies, such as model predictive control (MPC) or adaptive control, can optimize the operation of the distillation column in real-time. These systems can adjust variables like flow rates and temperatures to maintain optimal thermodynamic efficiency under varying conditions.

Thermodynamic efficiency plays a crucial role in determining the sustainability and cost-effectiveness of distillation processes. Understanding the limitations imposed by the Carnot efficiency and addressing the causes of exergy destruction can guide the design and operation of more energy-efficient distillation columns. By optimizing column design, operating conditions, and integrating energy-saving technologies, significant improvements in thermodynamic efficiency can be achieved. With growing concerns over energy consumption and environmental sustainability, the optimization of thermodynamic efficiency in distillation is more important than ever for industries relying on this separation technique.

Distillation column design strategies: From simple to robust methodologies

Distillation, a critical process in chemical engineering, involves the separation of mixtures based on differences in the boiling points of their components. Historically, distillation column design has evolved from simplistic, heuristic approaches to more sophisticated, data-driven strategies that prioritize efficiency, energy conservation, and process optimization. The need for such advancement is driven by growing industrial demands for better resource utilization, reduced operational costs, and improved sustainability. Over the years, the design strategies have shifted toward integrating process intensification principles to minimize energy usage and enhance the performance of distillation operations (Skiborowski & Górak, 2022).

Traditional distillation design strategies: Simple and heuristic approaches

In the early stages of distillation column design, engineers relied heavily on empirical, heuristic methods. These approaches were often based on practical experience and established guidelines. The design would generally involve optimizing parameters such as tray or packing configurations, reflux ratios, and feed locations.

One of the most basic approaches to distillation design is the McCabe-Thiele method, which assumes a constant molar overflow and equilibrium stages, using a graphical method to determine the number of stages required for separation. This approach was useful for relatively simple systems but lacked the precision needed for more complex mixtures. The method assumes ideal conditions, which is rarely the case in real-world applications. Nonetheless, the McCabe-Thiele method remains a key starting point for many distillation designs, particularly for binary mixtures (Sharma et al., 2020).

Another heuristic approach involves using the Fenske equation to estimate the minimum number of stages required for a given separation. The equation is relatively simple and does not require detailed process simulation, but it assumes ideal conditions, such as constant relative volatility, and may not always represent real-world behavior accurately.

These heuristic methods also emphasize reflux ratio optimization, where the reflux ratio is the ratio of the amount of liquid returned to the column as reflux to the amount of distillate product removed. In traditional designs, increasing the reflux ratio typically leads to a higher product purity. However, this comes at the expense of increased energy consumption and operational complexity (Lei et al., 2021).

Challenges with traditional methods

Despite their widespread use, traditional methodologies have limitations that hinder their ability to handle complex mixtures and optimize energy consumption. One significant challenge is the inherent inefficiency in energy usage. In classical distillation designs, energy-intensive reboilers and condensers are often used to achieve the required separation, leading to high operational costs. Furthermore, the simple approaches do not account for factors such as column hydrodynamics, pressure drops, or energy integration, which can have a profound effect on the column's performance.

Another limitation is the difficulty in adapting to multicomponent mixtures or non-ideal systems. Traditional methods, which often work well for binary systems, become less accurate and less efficient when applied to systems with more than two components. The assumption of constant relative volatility also becomes unrealistic as the system becomes more complex.

The shift to robust design methodologies

As the limitations of traditional distillation design strategies became evident, engineers began exploring more advanced methodologies to improve efficiency and performance. These robust design strategies focus on integrating new technologies, such as DWC, heat-integrated distillation systems, and reactive distillation, as well as using more accurate modeling techniques and process intensification principles (Kooijman & Sorensen, 2022).

Divided wall columns (DWC)

DWC are an advanced design that integrates multiple distillation processes into a single column by using a vertical wall to create separate sections within the column. This design enables the simultaneous separation of multiple components, which is particularly useful for multicomponent mixtures. The DWC improves separation efficiency and reduces energy consumption by consolidating what would otherwise be multiple distillation columns into one.

The key advantage of a DWC is its ability to reduce the number of stages required for separation by utilizing internal reflux streams and minimizing the total column height. A study by Kiss and Smith (2020) showed that DWCs could reduce energy consumption by up to 50% compared to conventional distillation columns for certain mixtures. Additionally, a DWC can reduce capital and operational costs by eliminating the need for multiple columns and associated equipment such as reboilers and condensers.

The optimization of DWCs requires advanced simulation techniques to accurately predict the behavior of the internal streams and ensure that the column design operates as intended. This is where the shift to data-driven and simulation-based design strategies has played a crucial role. In recent years, the development of computational fluid dynamics (CFD) simulations and process optimization algorithms has allowed for the precise design of DWCs, accounting for factors such as pressure drops, fluid flow behavior, and heat integration (Kiss & Smith, 2020).

Heat-integrated distillation systems

Heat integration is another key principle in robust distillation column design. The goal of heat integration is to reduce the external energy required for the distillation process by recovering and reusing heat within the system. This is particularly important in large-scale operations, where energy costs are a significant portion of operational expenses.

In heat-integrated distillation systems, the heat from the reboiler is recovered and used to provide the necessary heat for other parts of the process, such as the column's preheating section or even other unit operations within the plant. Pinch analysis is a common technique used to optimize heat integration by identifying where energy can be recovered and used more efficiently.

An example of this integration is the thermally coupled distillation column, where multiple distillation columns share heat and mass flows between them. This design reduces energy consumption by leveraging the heat from one column to support another. A study by Skiborowski (2023) showed that thermally coupled distillation systems could reduce energy consumption by as much as 30% compared to conventional, single-column systems.

Reactive distillation

Reactive distillation is another advanced design methodology that combines both separation and chemical reaction in a single unit operation. This design is particularly useful for processes that involve equilibrium-limited reactions, such as esterification or transesterification. By integrating reaction and separation into a single unit, reactive distillation can improve reaction conversion rates, reduce the number of units required, and lower energy consumption (Kiss, 2014).

In a reactive distillation column, the reaction occurs within the column's stages, and the volatile products of the reaction are continuously removed through distillation. This constant removal of products shifts the equilibrium, promoting higher conversion rates. In addition, the energy used for distillation helps drive the reaction forward by maintaining the required temperature for the chemical process.

The use of reactive distillation can also reduce the need for external heat sources, as the reaction itself provides the necessary heat. This integration of reaction and separation has been shown to reduce energy consumption by as much as 40% compared to traditional processes that separate the products first and then carry out the reaction in separate reactors. The design of reactive distillation columns requires detailed knowledge of both the chemical reaction kinetics and the distillation dynamics, which is often achieved through the use of process simulation tools.

Quantitative data on energy and cost savings

A comparison of various distillation strategies reveals significant energy and cost savings in advanced designs. For example, the use of DWCs can reduce energy consumption by approximately 50%, as noted earlier. In terms of cost savings, Kiss (2014) demonstrated that the capital cost for a DWC could be up to 30% lower than for a conventional distillation column system, owing to the reduced need for auxiliary equipment.

Similarly, heat-integrated distillation systems can significantly reduce operating costs. A study by Kiss (2014) found that the implementation of heat-integrated distillation systems could reduce energy consumption by up to 35%, with additional savings on cooling water usage. In terms of operating costs, the reduction in energy consumption translates to lower fuel and electricity bills, which can represent a savings of several million dollars annually for large industrial plants.

Reactive distillation also provides substantial energy savings, particularly in processes involving slow reactions or equilibrium-limited reactions. For example, in the production of biodiesel, the integration of transesterification and distillation in a single column can reduce energy costs by approximately 40%, according to Dimian and Kiss (2019).

The evolution of distillation column design from simple, heuristic-based strategies to advanced, robust methodologies represents a significant leap toward greater process efficiency and sustainability. The use of DWCs, heat-integrated systems, and reactive distillation highlights the ongoing trend toward process intensification in the chemical industry. These advanced strategies not only enhance separation efficiency but also lead to substantial reductions in energy consumption, operational costs, and environmental impact. As the industry continues to push for more sustainable and cost-effective solutions, the integration of advanced design methodologies, along with the use of process simulation tools and data-driven optimization techniques, will be crucial in shaping the future of distillation technology (Kiss et al., 2018).

Performance indicators in a sustainability framework associated with the energy consumption of the separation process

The distillation process, while critical in the chemical, petrochemical, and pharmaceutical industries, is also a major contributor to energy consumption and environmental impact. Energy usage in distillation not only affects the operational costs but also has significant consequences for sustainability, particularly in the context of global efforts to reduce carbon emissions and improve energy efficiency. Sustainability frameworks are essential in evaluating distillation systems, using key performance indicators (KPIs) such as carbon footprint, energy intensity, waste heat recovery potential, and others to gauge their environmental and economic impact (Kiss & Ignat, 2013).

In the context of distillation, it is imperative to consider energy consumption through the lens of sustainability, where the goal is not only to optimize performance but also to minimize the adverse environmental effects, such as greenhouse gas emissions and energy waste. This section delves into key performance indicators (KPIs) that allow for the assessment of distillation processes within sustainability frameworks, specifically focusing on energy consumption. Furthermore, we explore how life cycle analysis (LCA) and energy return on investment (EROI) metrics help balance the trade-offs between energy efficiency and environmental impact in distillation processes, while linking these indicators to advanced distillation designs such as DWCs, heat-integrated systems, and reactive distillation (Kiss & Smith, 2020).

Energy consumption in distillation: A sustainability concern

Energy consumption in distillation is a significant operational cost. Distillation columns, which are used to separate mixtures based on differences in boiling points, typically require substantial energy input for both the reboiler and condenser. The reboiler, which provides the necessary heat to vaporize the most volatile components of the mixture, accounts for a large portion of the total energy consumption in the distillation process. In conventional distillation columns, the energy required for this process can be high, especially for systems requiring high purity products or complex separations.

For example, conventional distillation systems can consume between 20% and 40% of a plant's total energy budget depending on the number of stages, feed composition, and separation requirements. When considering large-scale industrial applications, this translates to significant operational costs and carbon emissions. The high energy demand makes distillation a key target for process optimization, where improving energy efficiency can lead to both cost savings and reduced environmental impact.

In terms of greenhouse gas emissions, the energy used for distillation often comes from fossil fuel-based sources, which contribute to the carbon footprint of the process. This makes the energy consumption in distillation not only an operational issue but also an environmental concern, as higher energy use typically leads to greater emissions of CO_2, methane, and other greenhouse gases. Reducing energy consumption through innovative design methodologies is, therefore, a critical aspect of sustainable distillation.

Key performance indicators for energy consumption in distillation

To assess the energy performance of distillation systems, several KPIs are employed within sustainability frameworks. These indicators help quantify the environmental impact and efficiency of the separation process, allowing engineers and plant operators to identify opportunities for improvement (Kiss & Smith, 2020).

Carbon footprint

The carbon footprint measures the total amount of greenhouse gases emitted during the distillation process, typically expressed in grams of CO_2 equivalent per unit of product. This is one of the most widely used performance indicators in sustainability assessments, as it directly links energy consumption to environmental impact.

Carbon footprint can be influenced by multiple factors, including the energy sources used (e.g., fossil fuels, renewable energy), the efficiency of heat integration, and the design of the distillation column. For instance, distillation processes powered by renewable energy (e.g., solar or wind power) will have a much lower carbon footprint compared to those relying on coal or natural gas.

Quantifying the carbon footprint of a distillation column is essential for understanding its contribution to a facility's overall emissions. Studies show that reducing the energy consumption in distillation can cut carbon emissions by as much as 30%–50% when adopting advanced designs like DWC, heat-integrated systems, or reactive distillation. This reduction is achieved by optimizing the heat exchange and energy recovery processes, as well as reducing the number of stages required to achieve the desired separation.

Energy intensity

Energy intensity is a critical indicator used to measure the amount of energy required to produce a unit of product, typically expressed as kilowatt-hour per unit of product (kWh/unit). This KPI helps evaluate the efficiency of energy usage within the distillation process.

For example, conventional distillation columns may have an energy intensity of 3–5 kWh per liter of distillate for some separations, whereas more advanced configurations such as DWCs or thermally coupled distillation systems can achieve energy intensities as low as 1–2 kWh per liter for similar separations. This reduction in energy intensity is a key benefit of implementing process intensification strategies (Skiborowski, 2023).

Advanced distillation systems that integrate heat recovery or combine multiple distillation tasks into a single column (such as with reactive distillation) can drastically reduce energy intensity. A study conducted by

Janković et al. (2024) indicated that using a heat-integrated distillation system could reduce energy intensity by approximately 25%–40% compared to conventional systems, depending on the specific application.

Waste heat recovery potential

Waste heat recovery potential is another vital performance indicator. In distillation, a significant amount of energy is lost as heat from the reboiler and condenser. Capturing and reusing this waste heat can greatly enhance the sustainability of the process by reducing the need for external heating and cooling.

Heat-integrated distillation systems and DWCs are specifically designed to maximize waste heat recovery. By recovering the heat from the reboiler and redirecting it to preheat the feed or maintain column temperatures, these systems reduce the energy required to achieve the desired separation. According to Kostakis et al. (2018), waste heat recovery systems can cut external heating requirements by up to 50%, leading to substantial energy savings.

Life Cycle Analysis (LCA)

Life Cycle Analysis (LCA) is a holistic approach to evaluating the environmental impact of distillation processes throughout their entire life cycle, from raw material extraction to end-of-life disposal. LCA takes into account the energy consumption, material usage, emissions, and waste generated at every stage of the process, providing a comprehensive understanding of the sustainability of the distillation system.

LCA is particularly useful for comparing different distillation designs or technologies, as it helps assess not only operational impacts but also the energy and material inputs required for construction, maintenance, and decommissioning. For instance, LCA studies of conventional distillation columns versus DWCs reveal that while the initial capital costs of DWCs may be higher, the energy savings and reduced carbon emissions during operation lead to a lower overall environmental impact over the life cycle of the system.

Incorporating LCA into distillation design allows for better decision-making, ensuring that sustainability considerations are integrated from the start of the design phase (Skiborowski, 2023).

Energy Return on Investment (EROI)

Energy Return on Investment (EROI) is another metric that can be used to assess the efficiency of energy consumption in distillation. It is a measure of the energy output relative to the energy input, typically expressed as a ratio of the energy delivered by the system to the energy required to run it.

For distillation processes, a high EROI indicates that the system is operating efficiently, with minimal energy losses. Advanced distillation systems, such as heat-integrated columns or reactive distillation, tend to have higher EROI values because they recover and reuse energy more effectively. For example, EROI values for heat-integrated systems are often reported to be around 4:1 to 6:1, meaning that for every unit of energy input, the system delivers 4–6 times more energy in the form of separated products.

In contrast, conventional distillation columns typically exhibit lower EROI values, especially if they do not incorporate heat recovery or advanced process optimization techniques. The use of process intensification strategies can, therefore, lead to a higher EROI and greater overall energy efficiency (Skiborowski, 2023).

Integrating sustainability metrics with advanced distillation designs

When sustainability metrics such as carbon footprint, energy intensity, waste heat recovery, LCA, and EROI are applied to advanced distillation designs, such as DWCs, heat-integrated distillation systems, and reactive distillation, the advantages become even more pronounced (Kiss, 2014).

- DWC significantly reduce the energy required for multicomponent separations, leading to lower energy intensity and carbon footprint. The reduction in the number of stages and the heat integration within the column results in lower operational costs and energy consumption. These systems can improve energy efficiency by up to 50% and cut CO_2 emissions by a similar margin, according to recent studies.
- Heat-integrated distillation systems optimize energy usage by recovering waste heat and reusing it within the system, reducing the need for external heating and cooling. This can reduce overall energy consumption by up to 40%, leading to lower energy intensity and carbon emissions.
- Reactive distillation combines reaction and separation, reducing the need for multiple unit operations and optimizing both chemical reaction efficiency and energy usage. For processes such as biodiesel production, reactive distillation has been shown to reduce energy consumption by as much as 30% while increasing reaction rates.

The sustainability of distillation processes is becoming increasingly important in a world where energy efficiency and environmental impact are crucial factors in industrial decision-making. The adoption of performance indicators such as carbon footprint, energy intensity, and energy return on investment (EROI) allows for a more comprehensive evaluation of distillation systems. Moreover, the integration of advanced technologies, such as DWCs, heat-integrated distillation systems, and reactive distillation, offers significant improvements in energy efficiency and reductions in environmental impacts.

References

Altay, Ö., Aydın, Ö. K., Öner, H. B., Filiz, Ö., & Ertekin, F. K. (2025). Principles of distillation process. *Mass Transfer Operations in the Food Industry*, 45–81.

Cui, C., Qi, M., Zhang, X., Sun, J., Li, Q., Kiss, A. A., & Lee, M. (2024). Electrification of distillation for decarbonization: An overview and perspective. *Renewable and Sustainable Energy Reviews, 199*, Article 114522.

Dimian, A. C., & Kiss, A. A. (2019). Eco-efficient processes for biodiesel production from waste lipids. *Journal of Cleaner Production, 239*, Article 118073.

Halvorsen, I. J., & Skogestad, S. (2011). Energy efficient distillation. *Journal of Natural Gas Science and Engineering, 3*(4), 571–580.

Janković, T., Straathof, A. J., & Kiss, A. A. (2024). Process systems engineering perspectives on eco-efficient downstream processing of volatile biochemicals from fermentation. *Frontiers in Energy Research, 11*, Article 1340612.

Jobson, M. (2014). Energy considerations in distillation. In *Distillation* (pp. 225–270). Academic Press.

Kim, Y. H. (2016). Design and control of energy-efficient distillation columns. *Korean Journal of Chemical Engineering, 33*, 2513–2521.

Kiss, A. A. (2014). Distillation technology–still young and full of breakthrough opportunities. *Journal of Chemical Technology and Biotechnology, 89*(4), 479–498.

Kiss, A. A., & Ignat, R. M. (2013). Optimal economic design of an extractive distillation process for bioethanol dehydration. *Energy Technology, 1*(2-3), 166–170.

Kiss, A. A., Jobson, M., & Gao, X. (2018). Reactive distillation: Stepping up to the next level of process intensification. *Industrial and Engineering Chemistry Research, 58*(15), 5909–5918.

Kiss, A. A., Landaeta, S. J. F., & Ferreira, C. A. I. (2012). Towards energy efficient distillation technologies–Making the right choice. *Energy, 47*(1), 531–542.

Kiss, A. A., & Smith, R. (2020). Rethinking energy use in distillation processes for a more sustainable chemical industry. *Energy, 203*, Article 117788.

Kooijman, H. A., & Sorensen, E. (2022). Recent advances and future perspectives on more sustainable and energy efficient distillation processes. *Chemical Engineering Research and Design, 188*, 473–482.

Kostakis, V., Latoufis, K., Liarokapis, M., & Bauwens, M. (2018). The convergence of digital commons with local manufacturing from a degrowth perspective: Two illustrative cases. *Journal of Cleaner Production, 197*, 1684–1693.

Lei, Z., Dai, C., Chen, B., & Ding, Z. (2021). *Special distillation processes*. Elsevier.

Sharma, S., Kundu, A., Basu, S., Shetti, N. P., & Aminabhavi, T. M. (2020). Sustainable environmental management and related biofuel technologies. *Journal of Environmental Management, 273*, Article 111096.

Skiborowski, M. (2023). Synthesis and design methods for energy-efficient distillation processes. *Current Opinion in Chemical Engineering, 42*, Article 100985.

Skiborowski, M., & Górak, A. (2022). *Hybrid separation processes*. Berlin: De Gruyter.

Stichlmair, J. G., Klein, H., & Rehfeldt, S. (2021). *Distillation: Principles and practice*. John Wiley and Sons.

White, D. C. (2012). Optimize energy use in distillation. *Chemical Engineering Progress, 108*(3), 35–41.

Yang, K., Liu, S., He, C., Zhang, B., Chen, Q., & Pan, M. (2020). Improving energy saving of crude oil distillation units with optimal operations. *Journal of Cleaner Production, 263*, Article 121340.

Chapter 3

Thermal couplings: Motivation, physical implications in distillation columns

Distillation, a thermal separation method, is widely used to separate mixtures into their component fractions with a specific purity. This process exploits differences in the volatilities of components, which correlate with their boiling points, by applying and removing heat. Distillation is categorized as a physical separation process or unit operation and has diverse industrial applications, including crude oil refining (e.g., gasoline, diesel, kerosene), water purification and desalination, air separation (e.g., oxygen, nitrogen, argon), and the production of distilled beverages with high alcohol content (Kiss, 2013). Historically, distillation has undergone significant advancements due to its critical role in the petrochemical industry, making it one of the most important technologies in the global energy supply chain (Harmsen, 2010). Virtually all transportation fuels pass through at least one distillation column during their conversion from crude oil to usable fuel, with tens of thousands of columns operating worldwide.

In the face of declining fossil fuel reserves and the transition to renewable energy sources such as biomass, future transportation fuels are expected to include ethanol, methanol, or their derivatives. Producing these alternative fuels typically results in aqueous mixtures that require distillation to separate alcohols from water. Consequently, distillation remains the preferred separation method in the chemical process industry, underpinning the production of essential goods such as fuels, food, and raw materials for shelter and clothing.

Process intensification (PI) offers transformative innovations in equipment design and operation to improve process efficiency. PI aims to achieve dramatic reductions in plant volume—ideally by 100- to 1000-fold—by replacing conventional unit operations with novel, compact designs that often integrate multiple functions into a single unit (Boodhoo & Harvey, 2013). Over the past 2 decades, the definition of PI has evolved from simple miniaturization of equipment to include complex approaches, as described by Tian et al. (2018). These include combining multiple process tasks into a single unit (e.g., membrane reactors, reactive distillation), miniaturizing process equipment

Thermally Coupled Distillation Columns. https://doi.org/10.1016/B978-0-443-33114-5.00009-0

(e.g., microreactors), operating equipment periodically (e.g., simulated moving bed, pressure swing adsorption), and achieving tight process integration (e.g., dividing wall distillation). PI has emerged as a promising field that can enable paradigm shifts in the process industry by offering safer, more efficient operational methods (Segovia-Hernández & Bonilla-Petriciolet, 2015).

Distillation is inherently energy-intensive, accounting for approximately 40% of the total energy consumed in process industries, yet it offers a thermodynamic efficiency of only 5%–10% (Gómez-Castro et al., 2008). To address these limitations, heat integration techniques have been developed to enhance energy efficiency by generating internal thermal energy. Heat integration is categorized into internal and external approaches. Internal techniques include heat-integrated distillation columns and thermally coupled distillation sequences (TCDS), while external methods involve vapor recompression, bottoms flashing, closed-cycle compression, and absorption heat pumps (Kiss et al., 2016). This work focuses on TCDS.

The separation of multicomponent mixtures into three or more products is typically achieved using sequences of simple distillation columns with one feed stream and two product streams. However, complex distillation configurations, with or without thermal coupling, have demonstrated significant energy and exergy savings, stimulating scientific interest over the past 20 years. Numerous studies have examined the design and analysis of TCDS for ternary separations (e.g., Carlberg & Westerberg, 1989; Glinos & Malone, 1988; Rev et al., 2001; Tedder & Rudd, 1978; Yeomans & Grossmann, 2000). These studies show that thermally coupled configurations can achieve energy savings of up to 30% compared to conventional direct or indirect sequences, particularly for feeds with low or high intermediate-component content. For zeotropic ternary, quaternary, and multicomponent mixtures, various thermally coupled configurations have been proposed (Agrawal & Fidkowski, 1999; Blancarte-Palacios et al., 2003; Calzon-McConville et al., 2006; Hernández & Jiménez, 1996, 1999).

Over the years, several heuristics and procedures have been developed to identify optimal thermally coupled sequences for various mixtures. Notable contributions include studies by Dejanovic et al. (2010) and Asprion and Kaibel (2010), which provide comprehensive overviews of the research and implementation of TCDS, from early thermal coupling concepts to addressing practical challenges for successful industrial application. Despite the potential benefits and some reports of successful industrial adoption (Kaibel & Schoenmakers, 2002), TCDS implementation remains limited, primarily in Europe and Japan, where reliance on imported crude oil drives its adoption (Kim, 2006).

This chapter aims to introduce the basic principles of thermally coupled distillation, explore current strategies for designing optimal thermally coupled sequences.

Why do TCDS save energy? Remixing effect

Now, the multicomponent distillation applied to three component mixtures will be studied in order to exemplify thermodynamic inefficiencies of conventional distillation sequences.

When separating a three-product mixture using simple columns, there are only two possible sequences (see Fig. 3.1). Consider a system with the first characteristic of simple columns, a single feed is split into two products. As a first alternative to two simple columns, the possibilities shown in Fig. 3.2 can be considered. Here three products are taken from one column. The designs are in fact both feasible and cost-effective when compared to simple arrangements on a stand-alone basis (i.e., reboilers and condensers operating on utilities) for certain ranges of conditions.

If the feed is dominated by the middle product (typically more than 50% of the feed) and the heaviest product is present in small quantities (typically less than 5%), then the arrangements shown in Fig. 3.2a can be an attractive option. The heavy product must find its way down the column past the sidestream. Unless the heavy product has a small flow and the middle product a high flow, a reasonably pure middle product cannot be achieved. In these circumstances, the sidestream is usually taken as a vapor product to obtain a reasonably pure sidestream. A large relative volatility between the sidestream Product B and the bottom Product C is also necessary to obtain a high purity sidestream. A practical difficulty with this arrangement should, however, be noted. While it is straightforward to split a liquid flow in a column, it is far less straightforward to split a vapor flow as shown.

The flow of vapor up the column must somehow be restricted to allow a split to be made between the vapor that must continue up the column and the vapor sidestream. This can be achieved by the use of a baffle across the column diameter, or vapor tunnels in the tray downcomers. However, taking a vapor sidestream is more problematic than taking a liquid sidestream. If the

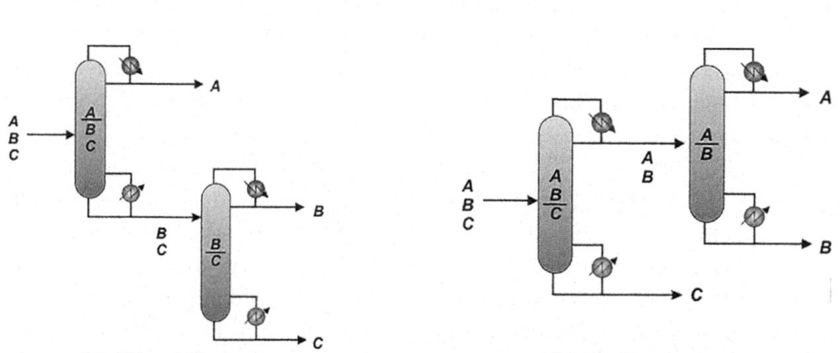

(a) Direct Sequence (b) Indirect Sequence

FIGURE 3.1 Conventional distillation sequences for three products.

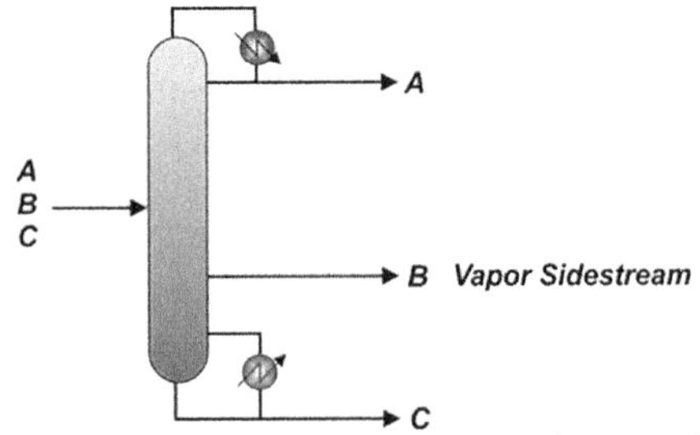

(a) More than 50% middle component and less than 5% heaviest component.

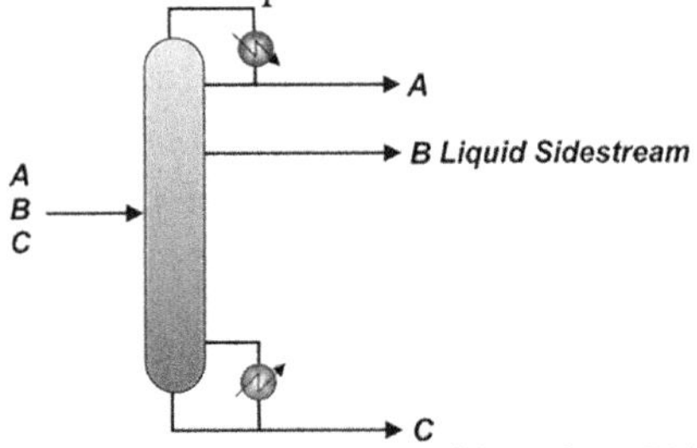

(b) More than 50% middle component and less than 5% lightest component.

FIGURE 3.2 Distillation column with three products.

feed is dominated by the middle product (typically more than 50%) and the lightest product is present in small quantities (typically less than 5%), then the arrangement shown in Fig. 3.2b can be an attractive option. This time the light product must find its way up the column past the sidestream. Again, unless the light product is a small flow and the middle product a high flow, a reasonably pure middle product cannot be achieved. This time the sidestream is taken as a liquid product to obtain a reasonably pure sidestream. A large relative volatility between the sidestream Product B and the overhead Product A is also necessary to obtain a high purity sidestream. In summary, single-column sidestream arrangements can be attractive when the middle product is in excess and one of the other components is present in only minor quantities. Thus, the sidestream column only applies to special circumstances for the feed composition. More generally applicable arrangements are possible by relaxing the restriction that separations must be between adjacent key components.

Consider a three-product separation as in Fig. 3.3a in which the lightest and heaviest components are chosen to be the key separation in the first column. Two further columns are required to produce pure products, Fig. 3.3a. This arrangement is known as distributed distillation or sloppy distillation. The distillation sequence provides parallel flow paths for the separation of a product. At first sight, the arrangement in Fig. 3.3a seems to be inefficient in the use of equipment in that it requires three columns instead of two, with the bottoms and overheads of the second and third columns both producing pure B.

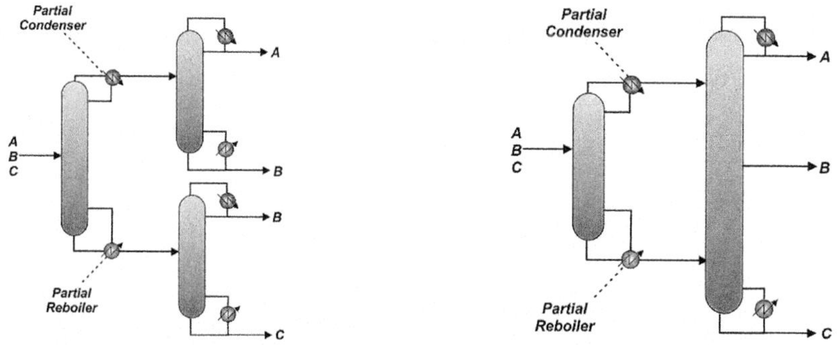

(a) Distributed distillation for nonadjacent key components

(b) Prefractionator Arrangement

FIGURE 3.3 Configuration with prefractionator.

However, it can be a useful arrangement in some circumstances. In the new design, the three columns can, in principle, all be operated at different pressures.

Also, the distribution of the middle Product B between the second and third columns is an additional degree of freedom in the design. The additional freedom to vary the pressures and the distribution of the middle product gives significant extra freedom to vary the loads and levels at which the heat is added to or rejected from the distillation. This might mean that the reboilers and condensers can be matched more cost-effectively against utilities, or heat integrated more effectively. If the second and third columns in Fig. 3.3a are operated at the same pressure, then the second and third columns could simply be connected, and the middle product taken as a sidestream as shown in Fig. 3.3b. The arrangement in Fig. 3.3b is known as a prefractionator arrangement. Note that the first column in Fig. 3.3b, the prefractionator, has a partial condenser to reduce the overall energy consumption. Comparing the distributed distillation in Fig. 3.3a and the prefractionator arrangement in Fig. 3.3b with the conventional arrangements in Fig. 3.1, the distributed and prefractionator arrangements typically require about 20%–30% less energy than conventional arrangements for the same separation duty. The reason for this difference is rooted in the fact that the distributed distillation and prefractionator arrangements are fundamentally thermodynamically more efficient than a simple arrangement. Based on the previous information, an important question arises, why does the prefractionator arrangement save energy? In order to respond this question, consider the sequence of simple columns shown in Fig. 3.4. However, moving further down the column, the composition of Component B decreases again as the composition of the less-volatile Component C increases.

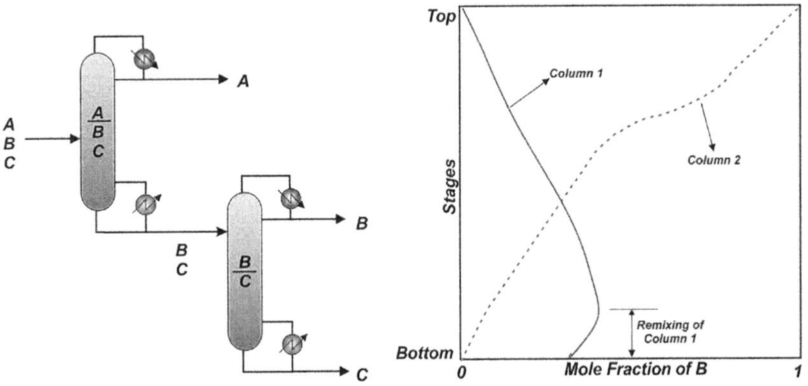

FIGURE 3.4 Composition profiles in direct sequence.

Thus, the composition of B reaches a peak only to be remixed subsequently. Similarly, with the first column in the indirect sequence, the composition of Component B first increases above the feed as the less-volatile Component C decreases. It reaches a maximum only to decrease as the more volatile Component A increases. Again, the composition of Component B reaches a peak only to be remixed. This remixing that occurs in both sequences of simple distillation columns affect the separation, thus it is considered the main source of inefficiency in multicomponent separation. Other potential sources of inefficiencies are the inherent remixing phenomena in condensers and reboilers and the differences or mismatches between the composition of the column feed and the composition at the feed stage, even after optimizing the location of this tray, which promotes the remixing effects. In theory, for binary distillation in a simple column, a good match can be found between the feed composition and the feed stage. However, because the changes from stage to stage are finite, an exact match is not always possible. For multicomponent distillation in a simple column, except under extreme circumstances, it is not possible to match the feed composition and the feed stage. The solution to eliminate or mitigate the remixing effect in multicomponent distillation systems was proposed by Petlyuk (1965). Consider, again the first column of Fig. 3.4, as previously mentioned, this column has a maximum concentration of the intermediate volatility component and then this concentration decreases. Petlyuk et al. proposed that if instead of forcing this first column to completely separate the most volatile component, the middle component is distributed smoothly throughout the first column as in a prefractionation column (see Fig. 3.3b), it is possible to guarantee a good match between the feed stages and the compositions in the main column, which reduces the remixing losses in these trays. Additionally, a high concentration of the middle component can be obtained in a sidestream in the main column, by the reduction of remixing losses as shown in Fig. 3.5. The elimination of remixing losses in the

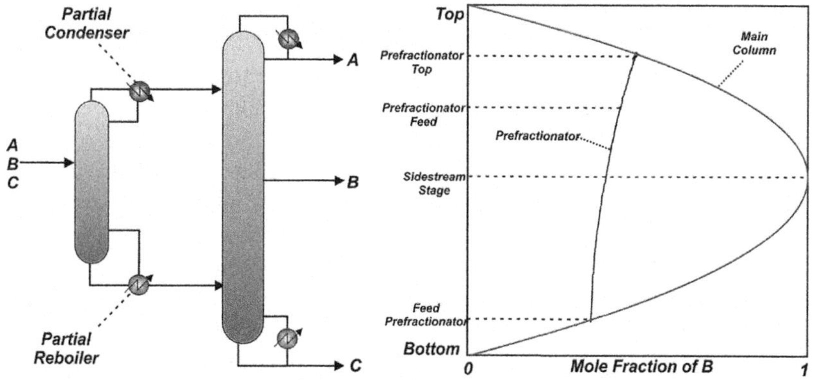

FIGURE 3.5 Remixing effect.

prefractionator arrangement means that it is inherently more efficient than an arrangement using simple columns. The same basic arguments apply to both distributed distillation and prefractionator arrangements, with the additional degree of freedom in the case of distributed distillation to vary the pressures of the second and third columns independently.

Thermally coupled columns

An important restriction of simple columns mentioned earlier was that they should have a reboiler and a condenser. However, it is possible to use material flows to provide some of the necessary heat transfer by direct contact, which also mitigates the remixing phenomena caused by the condensers and reboilers. This transfer of heat via direct contact is known as thermal coupling.

First consider thermal coupling of the simple direct sequence shown in Fig. 3.1. Fig. 3.6b shows a thermally coupled direct sequence. The reboiler of the first column is replaced by a thermal coupling. Liquid from the bottom of the first column is transferred to the second as before, but now the vapor required by the first column is supplied by the second column, instead of a reboiler on the first column. The four column sections are marked as 1, 2, 3, and 4 in Fig. 3.6a. It is important to mention that a section is a column part among condenser and feed or sidestream, reboiler and feed or sidestream, or between two streams. In Fig. 3.6c, the four column sections from Fig. 3.6b are rearranged to form a side-rectifier arrangement, this configuration is called thermally coupled equivalent. There is a practical difficulty in engineering a side-rectifier arrangement. This is the difficulty in taking a vapor sidestream from the main column. As already noted for vapor sidestream columns, while it is straightforward to split a liquid flow in a column, it is not straightforward to split a vapor flow as required in Fig. 3.6c. This problem can be avoided by constructing the side-rectifier in a single shell with a partition wall as shown in

Fig. 3.6d. The partition wall in Fig. 3.6d should be insulated to avoid heat transfer across the wall as different separations are carried out on each side of the wall and the temperatures on each side will differ. Heat transfer across the wall will have an overall detrimental effect on column performance.

Similarly, Fig. 3.7a shows an indirect sequence and Fig. 3.7b the corresponding thermally coupled indirect sequence. The condenser of the first column is replaced by thermal coupling. The four column sections are again marked as 1, 2, 3, and 4 in Fig. 3.7b. In Fig. 3.7c, the four column sections are rearranged to form a side-stripper arrangement. As with the side-rectifier, the side-stripper can be arranged in a single shell with a partition wall, as shown in Fig. 3.7d. Again, the partition wall should be insulated, otherwise heat transfer across the wall will have an overall detrimental effect on the separation. Both the side-rectifier and side-stripper arrangements have been shown to reduce the energy consumption compared to simple two-column arrangements. This results from reduced mixing losses in the first (main) column. As with the first column of the simple sequence, a peak in composition occurs with the middle product. But now, advantage of the peak is taken by transferring material to the side-rectifier or side-stripper.

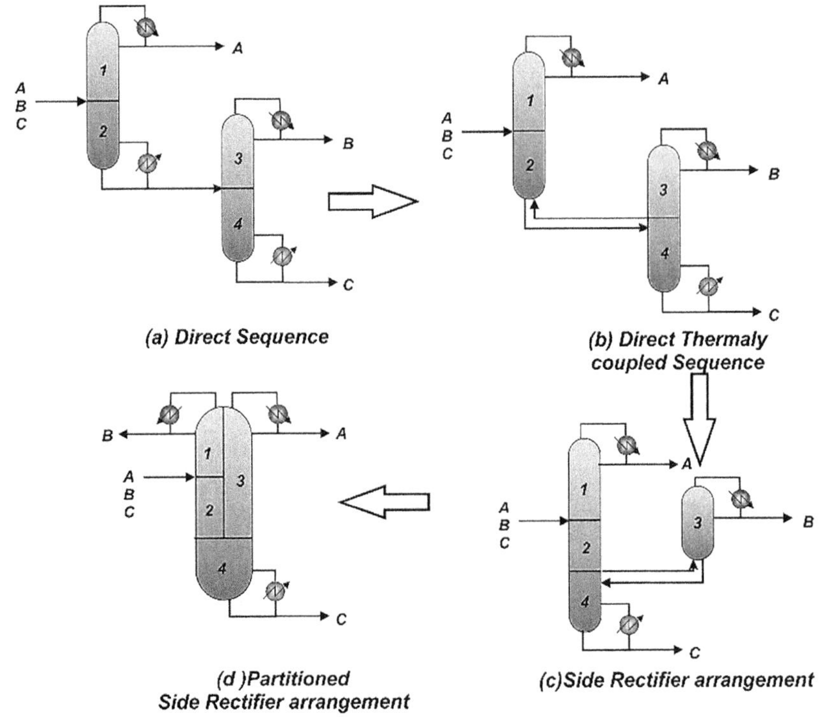

FIGURE 3.6 Thermally coupling in direct sequence.

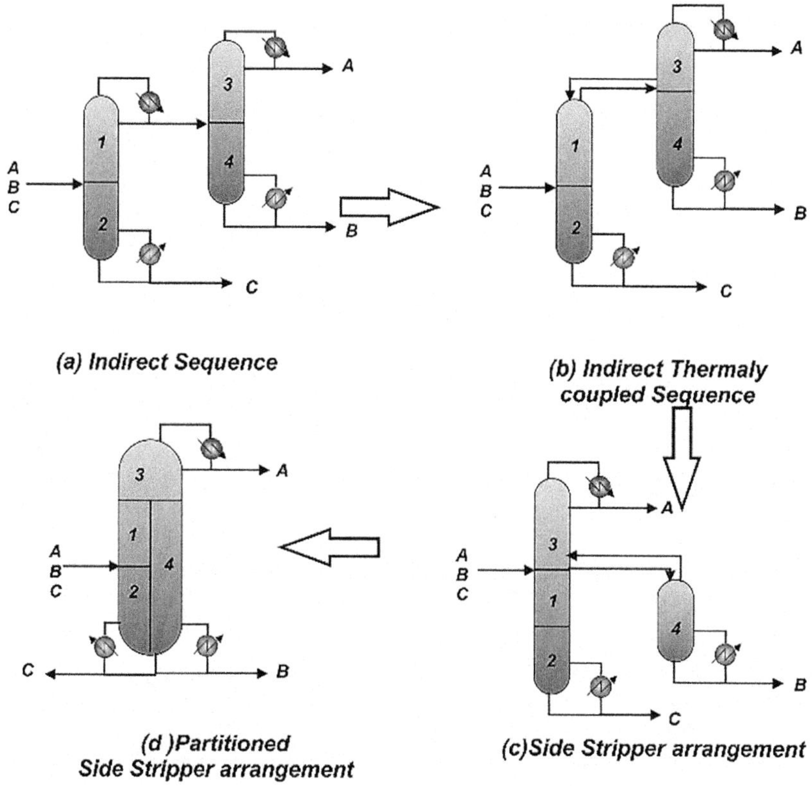

FIGURE 3.7 Thermally coupling in indirect sequence.

Now consider thermal coupling of the prefractionator arrangement from Fig. 3.3b. Fig. 3.8a shows a prefractionator arrangement with partial condenser and reboiler on the prefractionator.

Fig. 3.8b shows the equivalent thermally coupled prefractionator arrangement, sometimes known as the Petlyuk column. To make the two arrangements in Fig. 3.8a and b equivalent, the thermally coupled prefractionator requires extra plates to substitute for the prefractionator condenser and reboiler. The prefractionator arrangement in Fig. 3.8a and the thermally coupled prefractionator (Petlyuk column) in Fig. 3.8b are almost the same in terms of total heating and cooling duties. There are differences between the vapor and liquid flows in the top and bottom sections of the main column of the designs in Fig. 3.8a and b, resulting from the presence of the partial reboiler and condenser in Fig. 3.8a. However, although the total heating and cooling duties are almost the same, there are greater differences in the temperatures at which the heat is supplied and rejected. In the case of the prefractionator in Fig. 3.8a, the heat load is supplied at two points and two

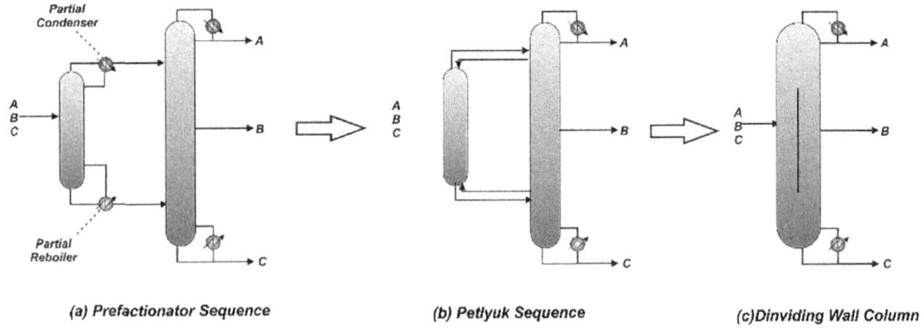

FIGURE 3.8 Thermally coupling in 3 columns arrangement.

different temperatures and rejected from two points and at two different temperatures. Fig. 3.8c shows an alternative configuration for the thermally coupled prefractionator that uses a single shell with a vertical partition dividing the central section of the shell into two parts, known as a dividing-wall column or partition column. The arrangements in Fig. 3.8b and c are equivalent if there is no heat transfer across the partition. As with side-rectifiers and side-strippers, the partition wall should be insulated to avoid heat transfer across the wall as different separations are carried out on each side of the wall and the temperatures on each side will differ. Heat transfer across the wall will have an overall detrimental effect on column performance. Partition columns offer several advantages over conventional arrangements: a) Various studies have compared the thermally coupled arrangement in Fig. 3.8b and c, with a conventional arrangement using simple columns on a stand-alone basis. These studies show that the prefractionator arrangement in Fig. 3.8 requires typically to 30% less energy than the best conventional arrangement using simple columns (Fig. 3.1). b) In addition, the partition column in Fig. 3.8c requires typically 20%–30% less capital cost than a two-column arrangement of simple columns. c) The partition column has one further advantage over the conventional arrangements in Fig. 3.1. In partitioned columns, the material is only reboiled once and its residence time in the high-temperature zones is minimized. This can be important if distilling heat sensitive materials.

Design of thermally coupled columns and effect of interconnection mass flows

During several years, the industrial implementation of thermally coupled systems was limited due to the lack of design procedures, in fact the development of a suitable design method was one important challenge for engineers and researchers. The main problem associated with the design of thermally coupled systems is their inherently greater complexity. This superior

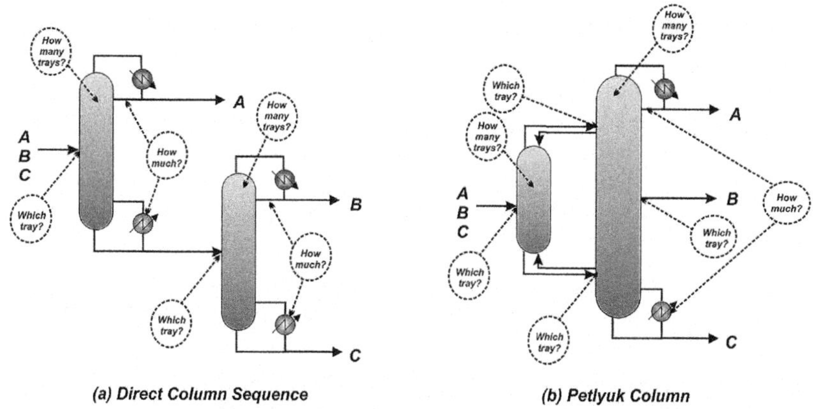

FIGURE 3.9 Degrees of freedom for conventional and Petlyuk arrangements.

complexity is directly related with many more degrees of freedom in contrast with the conventional distillation schemes (Dünnebier & Pantelides, 1999). Fig. 3.9 shows a comparison of the degrees of freedom for a conventional distillation and Petlyuk arrangements.

Most of the researchers have been focused on the development of short cut methods with the aim to get a fast screening of the possible alternatives and a good initialization of the degrees of freedom. However, most of the design methods have a restricted application and they usually solve de design problem assuming some supplications such as constant relative volatility or constant molal overflow. The solution to the design problem was provided by Hernández and Jiménez (1996), they suggested to use the design of conventional distillation systems to generate the thermally coupled arrangements. The design methodologies for conventional distillation systems are well known, these schemes are easily designed using conventional techniques such as Fenske- Underwood-Gilliland which provides the number or stages, feed stages, reflux ratios and reboiler duties to initialize the degrees of freedom. In this sense, once the conventional systems have been designed Hernández and Jiménez (1996) proposed to divide the column in sections in order to generate the design of the thermally coupled arrangements. It is important to mention that a section is a column part between a condenser or reboiler and feed or sidestream (see Figs. 3.6–3.7). Once the column sections have been identified and due to utilization of the liquid and vapor interconnection streams, it is quite simple to generate the thermodynamic equivalent arrangements moving the section between columns ash shown in Figs. 3.6 and 3.7, for this reason the method proposed by Hernández and Jiménez (1996) is also called moving section method.

Columns with lateral equipment

Despite this method has been considered one of the most successful for designing thermally coupled systems it does not provide an initial value for the interconnecting liquid or vapor streams, hence, a particular question arises about of how to specify or optimize those streams. Several previous works have determined that the interconnecting flows are one of the most important design variables because their values have direct effects on the energy requirements. Therefore, bad values for those parameters can provoke poor energy savings for thermally coupled arrangements or even more energy consumption compared to their conventional counterparts (Blancarte-Palacios et al., 2003; Dünnebier & Pantelides, 1999; Hernández et al., 2003). As a representative case, the effect of the interconnecting liquid on energy requirements for an indirect thermally coupled distillation scheme used to separate a equimolar mixture of n-pentane, n-hexane- n heptane is shown in Fig. 3.10 (Hernández et al., 2003). Note the presence of a minimum energy consumption for an interconnecting liquid flow around 25 kmol/hour, according to the data provided by Hernández et al., 2003, this minimum energy consumption corresponding to an energy saving of 21% in comparison with the indirect conventional sequence for separating the same mixture. In order to start the optimization procedure of the interconnection liquid showed in Fig. 3.10, the reflux rate of column C1 in the conventional indirect arrangement can be an initial value. In this sense, good initial values for the interconnecting streams can be obtained from the conventional schemes analyzing the flow rates or reflux ratios of the streams that connect both columns; these values can be easily calculated or obtained in simulation software.

Columns with total integration

Similarly, the interconnecting streams a strong effect on the energy consumption for the Petlyuk arrangements, however, due to presence of two pair of the interconnecting streams, at least one liquid and one vapor streams must

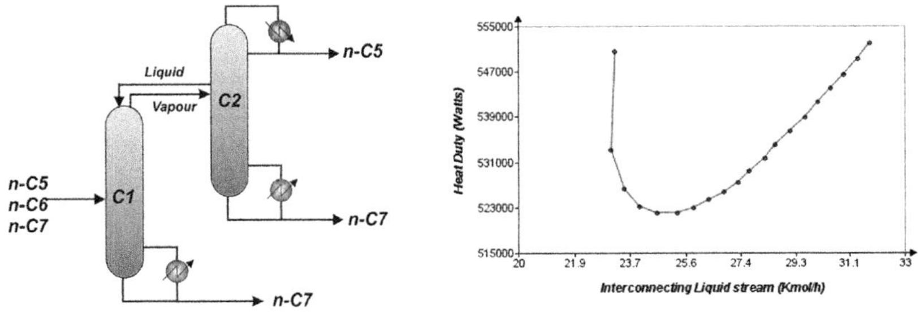

FIGURE 3.10 Effect of interconnecting flows on the energy consumption for an indirect thermally coupled sequence.

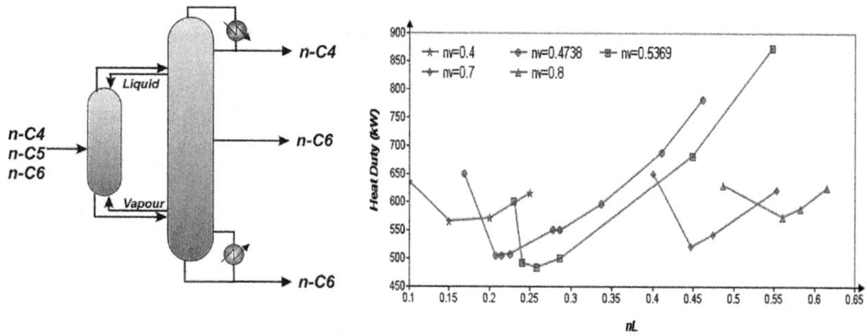

FIGURE 3.11 Effect of interconnecting flows on the energy consumption for an Petlyuk sequence.

be specified generating a three-dimensional surface. In the case of Petlyuk, column could be useful working with vapor or liquid fractions for interconnecting streams. The vapor fractions can be obtained from the conventional distributed scheme using the internal vapor and liquid flows of the stages where the thermally couplings are implemented (see Fig. 3.11). These internal mass flows can be obtained from simulation data (Hernández & Jiménez, 1999). The vapor fraction can be calculated using the following equations:

Where V_B is the interconnecting vapor flow, V_{NS2} is the internal vapor flow of the stage NS2, V_L is the interconnecting liquid flow, V_{NR} is the internal liquid flow of the stage NR. Based on the previous work reported by Hernández & Jiménez, 1999, it is quite simple to set firstly the vapor fraction based on data from conventional arrangements. At the same time, Hernández and Jiménez (1999) found that usually the liquid fraction value is lower than the vapor fraction, which significantly reduces the search space in order to achieve a Petlyuk scheme with minimum energy consumption. Fig. 3.11 shows the effect of the interconnecting liquid and vapor streams on the energy consumption of a Petlyuk column used to separate an equimolar mixture of n-butane, n-pentane, n-hexane (Hernández & Jiménez, 1999).

Bidirectionality of Petlyuk column: Generation of a dividing wall column

The method proposed by Hernandez and Jimenez provides a powerful tool to generate different thermally coupled designs considering only the different sections of distillation columns, this can be achieved because, this method takes advantage of flow direction of interconnecting streams and the internal liquid and vapor flows. Formally, two configurations can be considered thermodynamically equivalent if one arrangement can be rearrangement simply by moving sections from another scheme, which generates a changing on direction of interconnecting streams (see Figs. 3.6 and 3.7). In these cases, the change on interconnecting flow directions flows direction does affect thermodynamic advantages of thermally coupling, this can be easily corroborated analyzing an enthalpy diagram for the thermally schemes. Firstly, consider a single column in which the flow of the feed stream enters the column first as saturated liquid and then as superheated steam. The enthalpy -temperature diagrams for this case are shown in Fig. 3.12. Based on the enthalpy diagrams, the energy requirements for a saturated liquid feed stream are greater than for an overheating vapor feed, because the reboiler requires overcoming the latent heat barrier in order to vaporize the feed stream.

Now consider the direct distillation sequences and their respective thermally coupled arrangement, the enthalpy diagrams for both sequences are shown in Fig. 3.13. Note in Fig. 3.13, that because the interconnection streams contain liquid and vapor flows at their respective saturation point, the compiled behavior of the two interconnecting streams that generate the thermal coupling among both columns, is equivalent to a net current of superheat steam, which reduces the energy requirements used for heating in the thermally coupled schemes (Carlberg & Westerberg, 1989a). This compiled behavior of feed streams persists in the thermally equivalent columns and it is independent of the direction of the interconnecting liquid and vapor flows, therefore, it is also called bidirectionality. A similar analysis can be performed for the indirect arrangements (Caballero & Grossmann, 2013).

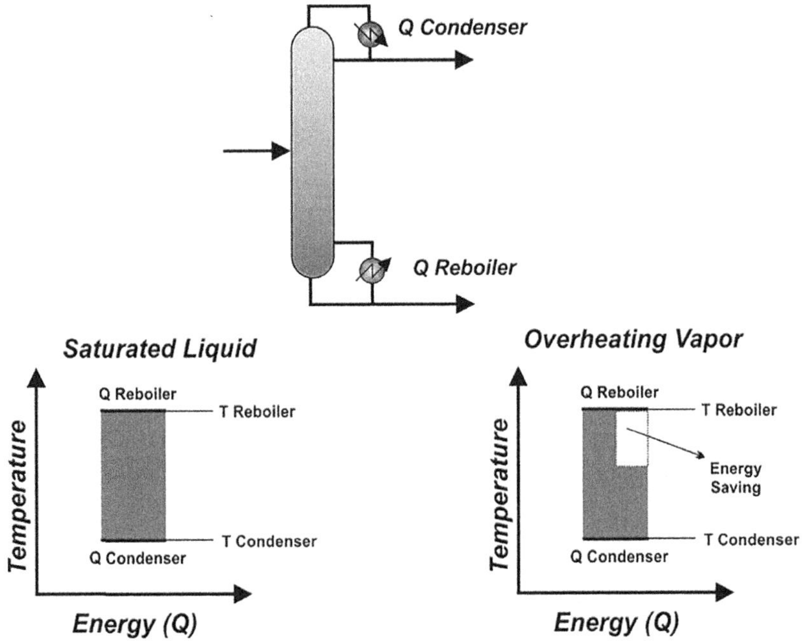

FIGURE 3.12 Enthalpy diagrams for saturated liquid and overheating vapor feeds.

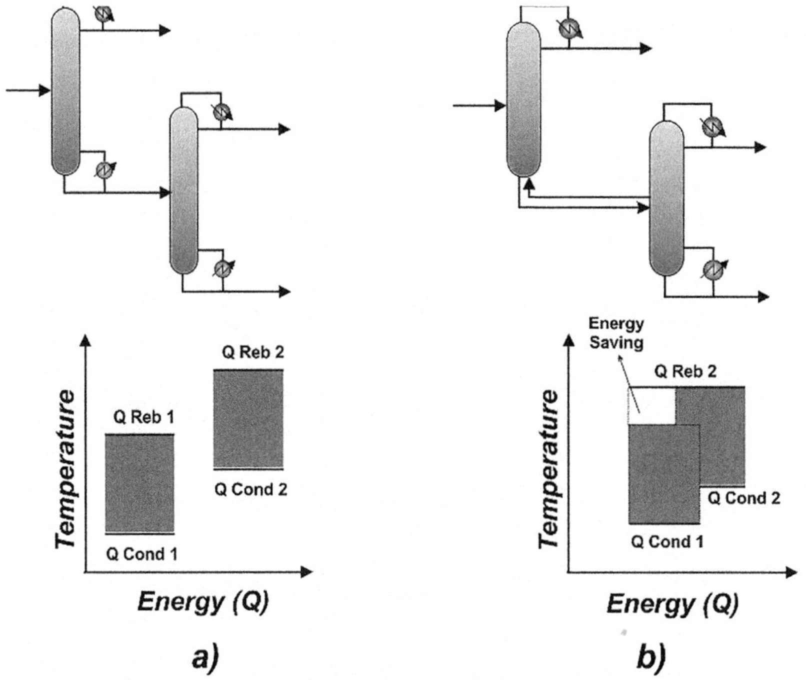

FIGURE 3.13 Enthalpy diagrams (a) Conventional direct sequence (b) Direct thermally coupled sequence.

Carlberg and Westerberg (1989b) demonstrated that the energy savings in the Petlyuk column can be represented by a system where the direct and indirect thermally coupled schemes occur simultaneously in the prefractionator column, for this reason the Petlyuk arrangement is also known as distillation with fully thermally coupled. Therefore, the Petlyuk column can be represented by an equivalent three-column scheme however in order to represent those three columns alternative, it is necessary to adjust the energy provided in the boiler and the condenser energy load to match the flows at the stage of extraction of intermediate volatility component. Fig. 3.14 exemplifies this equivalence.

Because the Petlyuk column only uses one condenser and reboiler instead of four heat exchanges of the conventional distillation sequences, it is possible to obtain important investment savings. Finally, the bidirectional streams can be rearranged for generating a diving wall column, where the prefactionator and the main column are integrated in a single shell and divided by an internal wall (see Fig. 3.15).

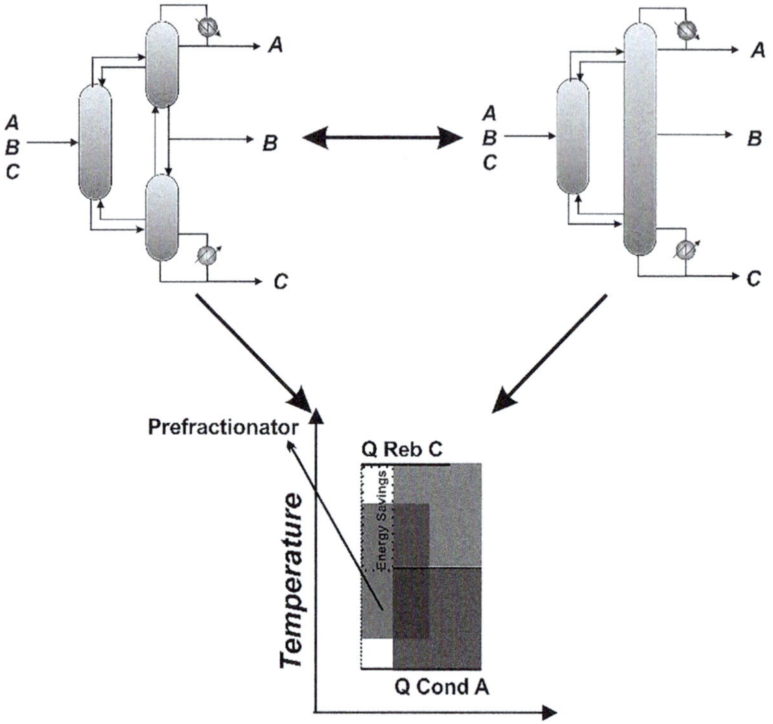

FIGURE 3.14 Enthalpy diagrams for a Petlyuk column and an equivalent three columns arrangement.

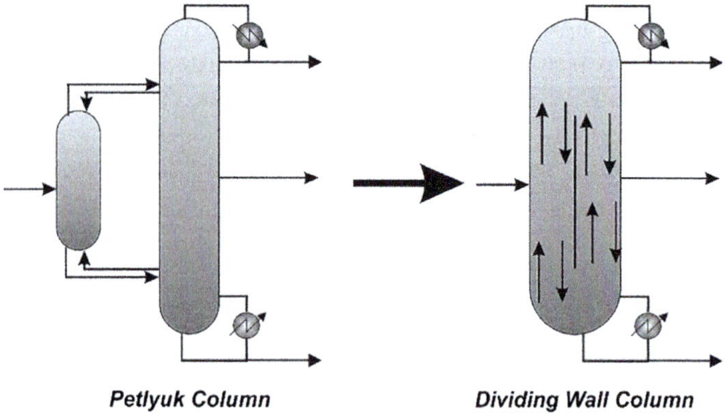

FIGURE 3.15 Enthalpy diagrams for a Petlyuk column and an equivalent three columns arrangement.

Implementation in mixtures with 4 or more components

The thermally couped distillation schemes can be also implemented for systems with four or more components in order to reduce the energy consumption of the separation. However, when the number of components is higher than three components the set of possible thermally coupled schemes increase dramatically, most of those alternatives do not appear in three component systems. For this reason, some important structural considerations should be taken into account, in order to generate as much as possible of alternatives and determinate the best scheme. In conventional distillation arrangements, N components can be separated using N-1 columns, condensers and reboilers, these N-1 columns are equivalent to 2(N-1) column sections (Agrawal, 1999, 2000). Based on the aforementioned information, it is possible to replace condenser and reboilers by interconnection columns, without modifying the process structure. In this case, the number of heat exchangers is the same than the number of components, as is shown in Fig. 3.16.

The replacement of condensers and reboilers for the unassociated product streams generates direct and indirect thermally coupled schemes. Note that in this case the indirect and direct arrangements coexist in a single distillation sequence, which is very common in mixtures with 4 or more components. On the other hand, the thermally coupled sequence showed in Fig. 3.16 can be easily modified to generate a fully thermally coupled schemes, by replacing the heat exchangers associated with streams of intermediate volatility pure components by interconnecting streams. Nevertheless, in order to achieve the full thermally coupling and the desired purities, two extra column sections must be added for each exchanger removed. The number of stages of those new sections corresponds to the minimum theoretical stages required to separate the corresponding adjacent components. The minimum stages are calculated using the Fenske-Underwood-Gilliland method. Therefore, the minimum number of column sections required to generate a full thermally couped arrangement of mixtures with four or more components is 4N–6 (Agrawal, 1999). The addition of this extra stages is presented in Fig. 3.17 in red numbers. Note that although the minimum number of column section is

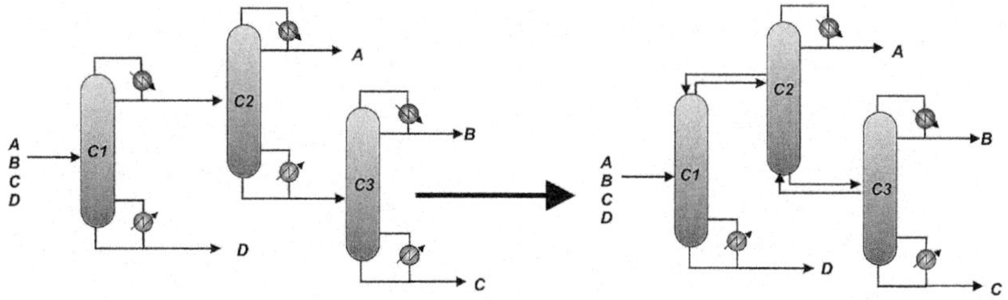

FIGURE 3.16 Thermally coupled arrangement for a four components separation.

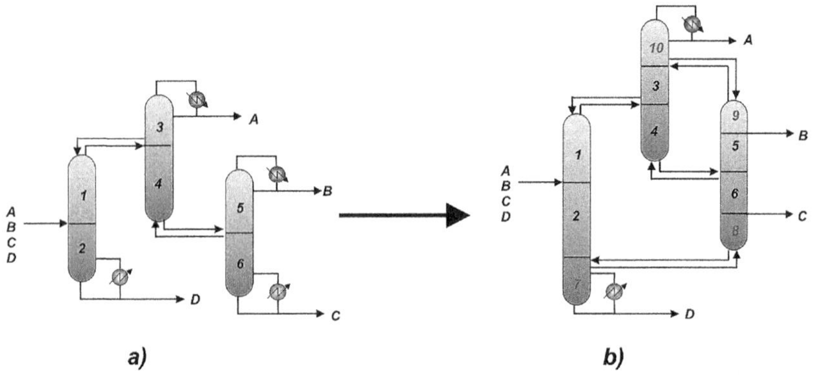

FIGURE 3.17 Generation of fully thermally coupled arrangements in a four components system (a) Thermally coupled scheme using six sections (b) Full thermally coupled scheme using 10 sections.

4N–6 (without considering possible internal walls shames), the number of columns required to perform the separation is kept at N-1 columns.

It is important to highlight, that thermo equivalent arrangements for systems with 4 or more components can be generated in analog way to three components systems using the moving section method. However, due to the numerous possible alternatives, many thermo equivalent configurations could be omitted with the methodology. Additionally, when a multicomponent thermally coupled arrangement is considered, not always is easy to discern whether two sequences are thermodynamically equivalent or not. For example, Fig. 3.18 shows two thermodynamically equivalent sequences, although both alternatives look as two different configurations. The structure of full thermally coupled sequence of Fig. 3.18a is called satellite column arrangement

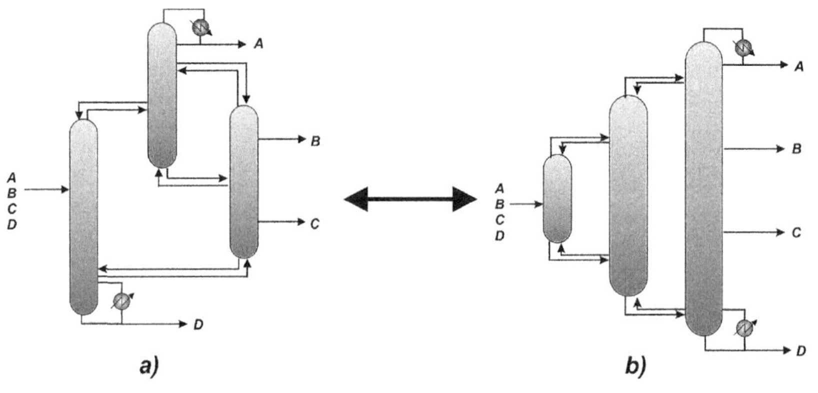

FIGURE 3.18 Thermally coupled equivalent configurations for a four components system (a) satellite column arrangement (b) sequential column arrangement.

whereas the scheme of Fig. 3.18b is known as sequential column arrangement or general thermally coupled system for four components (Agrawal., 2000; Sargent & Gaminibandara., 1976).

When two or more thermally coupled sequences are "thermodynamically equivalent" those configurations will have the same number of stages, column sections, products concentrations, energy requirements, whereas the main differences between those schemes are in the vapor and liquid flows between sections, and mainly in the order of separation among different components. The number of columns sections, stages, compositions, and energy requirements can be easily estimated using as reference the respective conventional distillation sequence. For this reason, a simple way of reducing or enclosing the number of possible thermally coupled configurations, and at the same time, selected the most adequate set of possible schemes, is selected the best conventional distillation sequence. The selection of the best conventional alternative is usually carried out taking into account a set of different metrics such as: minimum energy requirements, minimum cost, best controllability criteria, safety properties among others. However, the difficulty in selecting the best conventional alternative increases with the number of components involved in the mixture. Fig. 3.19 presents the superstructure of possible separation routes for three, four, five, and six-component mixtures. Components in a mixture are ranked according to their relative volatility; where A is the most volatile component and volatility decreases in successive order.

Based on Fig. 3.19 note that the number of possible separation alternatives increases with the number of components. The selection of a specific separation alternative depends on many factors such as: thermodynamic properties of mixture, components compositions, energy requirements, economic, control, safety, and environmental issues, among others. Once the best alternative is conventional sequence selected, the set of thermally coupled schemes is derivate. Different heuristic rules have been developed in order to achieve the best conventional configuration. In this sense, Agrawal (2003) proposed six points for selecting basic distillation configurations, these points are mentioned as follows:

1. The lightest component (A) is always recovered at the top of a distillation column and has an associated condenser. In contrast, the heaviest component is always recovered at the bottom of a distillation column and it has an associated reboiler.
2. A component of intermediate volatility may be recovered with an associated condenser or a reboiler or neither of the two.
3. Whenever a submixture is associated with a condenser or a reboiler, this submixture is always removed from the distillation column.
4. If a submixture is transferred from one distillation column to another distillation column, then both the sections emanating from it are needed and they exist in the distillation column receiving the submixture.

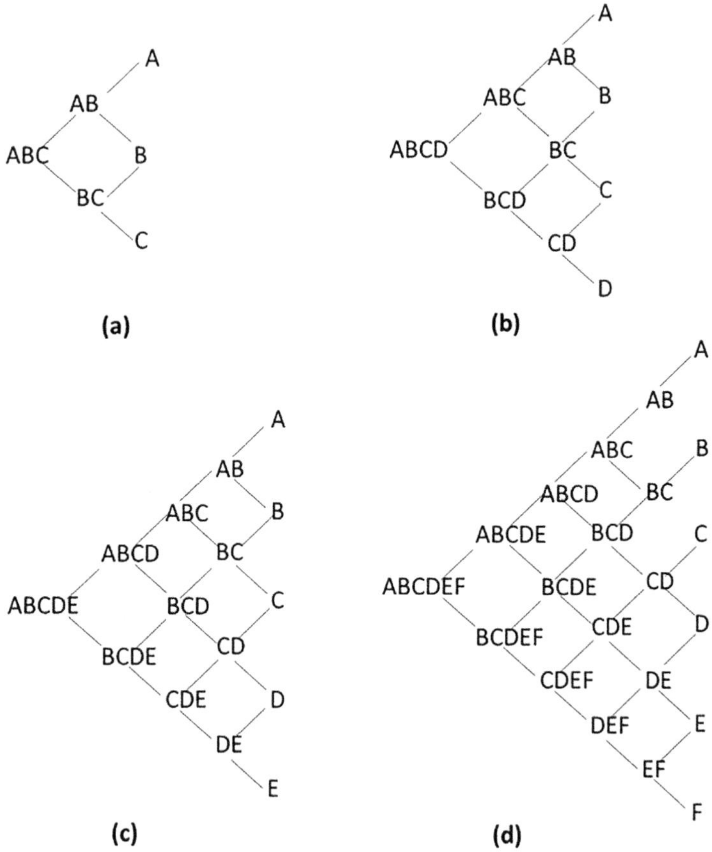

FIGURE 3.19 Possible separation alternatives for (a) three, (b) four, (c) five and (d) six-components.

5. When a submixture on the upper branch is transferred from one distillation column to another, then it has a condenser associated with it at the withdrawal location. In contrast, when a submixture on the lower branch is transferred from one distillation column to another then it has a reboiler associated with it at the remotion location. In addition, an internal submixture, which is neither on the upper nor the lower branch, can be transferred between the columns with or without an associated reboiler or condenser.
6. When a condenser is associated with an internal submixture at a depth m in a network, then transfers of all submixtures are eliminated at depths earlier than m that contain the internal submixture as the heavy subgroup within

the submixture. On the other hand, when a reboiler is associated with an internal submixture at depth m, transfer of all submixtures at depths earlier than m that contain the internal submixture as the light subgroup within the submixture are eliminated. Finally, for the recovery of an internal submixture at depth m from a distillation column without an associated reboiler or a condenser, there must be distillation sections both above and below the withdrawal point.

Industrial application of thermally coupled and divided wall column configurations

The first known application of thermally coupled distillation is due to Wright and Elizabeth (1949), who in 1949 patented an apparatus for separating air components, nowadays, this process is known as an internal wall column. Despite the thermally coupled schemes have been proposed since 1969 and provide important energy savings about 30%–50% in contrast to the conventional configurations, it has not been until the last years that the thermally coupled schemes started to be implemented with more frequency in the chemical industry. The reason is quite simple, and it is associated with the implementation of the interconnecting streams required to achieve energy reductions. These interconnecting streams raise the complexity of the thermally coupled arrangements. For many years the engineers and researchers thought that the energy savings were being achieved at the expense of designing a more difficult process to control. Due to this the controllability of thermally coupled processes started to be investigated, reaching the conclusion that the thermally coupled processes have similar control properties and, in many cases, they may even have better control properties than the conventional configurations.

Currently, the BASF company is the world leader in the implementation of thermally coupled and divided wall columns, with more than 50 processes operating these technologies for the year 2002 (Schultz et al., 2002; Yildirim et al., 2011). On other hand, the thermally couped columns the thermally coupled configurations have been applied mainly of a broad spectrum of mixtures such as: hydrocarbons, alcohols, aldehydes, ketones, acetals, amines, and others. Additionally, these technologies have been implemented in complex distillation systems e.g., azeotropic, extractive and reactive distillation (Yildirim et al., 2011). Finally, Fig. 3.20 shows the number of thermally coupled columns implemented over the years and the corresponding implementation predictions according to the data provided by Yildirim et al. (2011). Note, that for the year 2020 around 1100 thermally coupled should be implemented, however, these data are difficult to corroborate because the implementations of these technologies are commonly industrial secrets (Table 3.1).

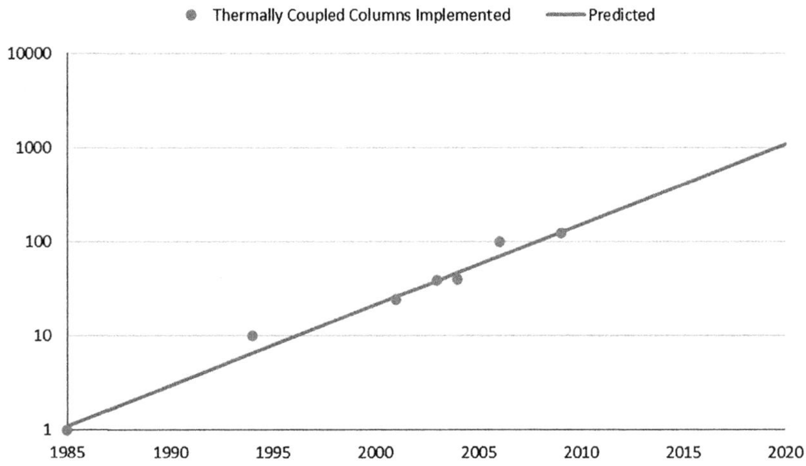

FIGURE 3.20 Industrial-scale implementation of thermally coupled distillation columns.

TABLE 3.1 Thermally coupled columns implemented in the industry (Yildirim et al., 2011).

Systems	Number of thermally coupled columns
Ternary mixture	116
Mixture with four or more components	2
Reactive systems	–
Azeotropic mixture	1
Extractive systems	2
Revamps	4

Conclusions

The energy uses in the chemical process industry (particularly related to distillation) is essential for improving the sustainability of the chemical sector. Some concluding remarks can be established.

1. Despite the maturity of distillation, it still presents an enormous scope to reduce its energy consumption by significantly improving its low thermodynamic efficiency using thermally couplings, for example.

2. Thermally coupled distillation technologies should not be considered in isolation as solution to a problem, but in the context of the overall system in order to find the optimal solution.
3. Optimization of TCDS requires all of the degrees of freedom to be manipulated simultaneously, but this is challenging.

References

Agrawal, R., & Fidkowski, Z. T. (1999). New thermally coupled schemes for ternary distillation. *AIChE Journal, 45*(3), 485–496.

Agrawal, R. (1999). More operable fully thermally coupled distillation column configurations for multicomponent distillation. *Chemical Engineering Research and Design, 77*(6), 543–553.

Agrawal, R. (2000). A method to draw fully thermally coupled distillation column configurations for multicomponent distillation. *Chemical Engineering Research and Design, 78*(3), 454–464.

Agrawal, R. (2003). Synthesis of multicomponent distillation column configurations. *AIChE Journal, 49*(2), 379–401.

Asprion, N., & Kaibel, G. (2010). Dividing wall columns: Fundamentals and recent advances. *Chemical Engineering and Processing: Process Intensification, 49*(2), 139–146.

Blancarte-Palacios, J. L., Bautista-Valdés, M. N., Hernández, S., Rico-Ramírez, V., & Jiménez, A. (2003). Energy-efficient designs of thermally coupled distillation sequences for four-component mixtures. *Industrial & Engineering Chemistry Research, 42*(21), 5157–5164.

Boodhoo, K., & Harvey, A. (Eds.). (2013). *Process intensification technologies for green chemistry: Engineering solutions for sustainable chemical processing*. John Wiley & Sons.

Caballero, J. A., & Grossmann, I. E. (2013). Synthesis of complex thermally coupled distillation systems including divided wall columns. *AIChE Journal, 59*(4), 1139–1159.

Calzon-McConville, C. J., Rosales-Zamora, M. B., Segovia-Hernández, J. G., Hernández, S., & Rico-Ramírez, V. (2006). Design and optimization of thermally coupled distillation schemes for the separation of multicomponent mixtures. *Industrial & Engineering Chemistry Research, 45*(2), 724–732.

Carlberg, N. A., & Westerberg, A. W. (1989a). Temperature-heat diagrams for complex columns. 2. Underwood's method for side strippers and enrichers. *Industrial & Engineering Chemistry Research, 28*(9), 1379–1386.

Carlberg, N. A., & Westerberg, A. W. (1989b). Temperature-heat diagrams for complex columns. 3. Underwood's method for the Petlyuk configuration. *Industrial & Engineering Chemistry Research, 28*(9), 1386–1397.

Dejanović, I., Matijašević, L., & Olujić, Ž. (2010). Dividing wall column—a breakthrough towards sustainable distilling. *Chemical Engineering and Processing: Process Intensification, 49*(6), 559–580.

Dünnebier, G., & Pantelides, C. C. (1999). Optimal design of thermally coupled distillation columns. *Industrial & Engineering Chemistry Research, 38*(1), 162–176.

Glinos, K., & Malone, M. F. (1988). Optimality regions for complex column alternatives in distillation systems. *Chemical Engineering Research and Design, 66*(3), 229–240.

Gómez-Castro, F. I., Segovia-Hernández, J. G., Hernandez, S., Gutiérrez-Antonio, C., & Briones-Ramírez, A. (2008). Dividing wall distillation columns: Optimization and control properties. *Chemical Engineering & Technology: Industrial Chemistry-Plant Equipment-Process Engineering-Biotechnology, 31*(9), 1246–1260.

Harmsen, J. (2010). Process intensification in the petrochemicals industry: Drivers and hurdles for commercial implementation. *Chemical Engineering and Processing: Process Intensification, 49*(1), 70–73.

Hernández, S., & Jiménez, A. (1996). Design of optimal thermally-coupled distillation systems using a dynamic model. *Chemical Engineering Research and Design, 74*(3), 357–362.

Hernández, S., & Jiménez, A. (1999). Design of energy-efficient Petlyuk systems. *Computers & Chemical Engineering, 23*(8), 1005–1010.

Hernández, S., Pereira-Pech, S., Jiménez, A., & Rico-Ramírez, V. (2003). Energy efficiency of an indirect thermally coupled distillation sequence. *Canadian Journal of Chemical Engineering, 81*(5), 1087–1091.

Kaibel, G., & Schoenmakers, H. (2002). Process synthesis and design in industrial practice. *Computer Aided Chemical Engineering, 10*, 9–22.

Kim, Y. H. (2006). A new fully thermally coupled distillation column with postfractionator. *Chemical Engineering and Processing: Process Intensification, 45*(4), 254–263.

Kiss, A. A. (2013). Novel applications of dividing-wall column technology to biofuel production processes. *Journal of Chemical Technology & Biotechnology, 88*(8), 1387–1404.

Kiss, A. A., Lange, J. P., Schuur, B., Brilman, D. W. F., van der Ham, A. G., & Kersten, S. R. (2016). Separation technology–Making a difference in biorefineries. *Biomass and Bioenergy, 95*, 296–309.

Petlyuk, F. B. (1965). Thermodynamically optimal method for separating multicomponent mixtures. *International Chemical Engineering, 5*, 555–561.

Rév, E., Emtir, M., Szitkai, Z., Mizsey, P., & Fonyo, Z. (2001). Energy savings of integrated and coupled distillation systems. *Computers & Chemical Engineering, 25*(1), 119–140.

Sargent, R. W. H., & Gaminibandara, K. (1976). *Optimum design of plate distillation columns.*

Schultz, M. A., Stewart, D. G., Harris, J. M., Rosenblum, S. P., & Shakur, M. S. (2002). Reduce costs with dividing-wall columns. *Chemical Engineering Progress, 98*(5), 64–71.

Segovia-Hernández, J. G., & Bonilla-Petriciolet, A. (2015). *Process intensication in chemical engineering*. Springer.

Tedder, D. W., & Rudd, D. F. (1978). Parametric studies in industrial distillation: Part I. Design comparisons. *AIChE Journal, 24*(2), 303–315.

Tian, Y., Demirel, S. E., Hasan, M. F., & Pistikopoulos, E. N. (2018). An overview of process systems engineering approaches for process intensification: State of the art. *Chemical Engineering and Processing-Process Intensification, 133*, 160–210.

Wright, R. O., & Elizabeth, N. J. (1949).

Yeomans, H., & Grossmann, I. E. (2000). Optimal design of complex distillation columns using rigorous tray-by-tray disjunctive programming models. *Industrial & Engineering Chemistry Research, 39*(11), 4326–4335.

Yildirim, Ö., Kiss, A. A., & Kenig, E. Y. (2011). Dividing wall columns in chemical process industry: A review on current activities. *Separation and Purification Technology, 80*(3), 403–417.

Chapter 4

Synthesis, design, and optimization strategies in the application of thermal couplings

Introduction

Distillation is one of the most important and widely used separation techniques in chemical engineering, renowned for its simplicity and effectiveness in separating liquid mixtures based on differences in volatility. Its application spans numerous industries, including petrochemicals, pharmaceuticals, and food processing, highlighting its versatility in producing high-purity products. However, as separation requirements grow more complex, particularly for multicomponent mixtures, traditional approaches to distillation face challenges that necessitate innovative solutions.

Historically, the design of distillation systems for multicomponent mixtures has been a central topic in chemical engineering research. The foundational work of Underwood and Fenske (Underwood, 1949) laid the groundwork for calculating minimum reflux ratios and determining the minimum number of theoretical stages required for binary separations. These methods were instrumental in advancing the understanding of distillation column behavior and were later extended to multicomponent systems. Gilliland's contributions added correlations for predicting the number of stages needed for a given separation, creating a bridge between the theoretical and practical aspects of distillation design. These early methodologies provided a fundamental understanding of the separation process and led to the development of systematic techniques to design simple column sequences for separating mixtures with multiple components (Seader et al., 2011).

As the number of components in a mixture increases, the design complexity of the distillation system grows exponentially. The number of feasible configurations, or separation sequences, can be vast due to the factorial nature of combinations in which components can be split across

Thermally Coupled Distillation Columns. https://doi.org/10.1016/B978-0-443-33114-5.00011-9

different columns. The challenge is further compounded when energy efficiency considerations are introduced. Thermal coupling between columns, a strategy that has gained traction for reducing energy consumption, adds an additional layer of complexity to the design problem. The introduction of thermal couplings can significantly reduce energy requirements by minimizing the number of reboilers and condensers, but it also expands the number of possible configurations, making the search for the optimal design a highly intricate process.

Various strategies have been proposed to tackle this complexity, focusing on generating and evaluating different distillation configurations to identify optimal sequences. Nath and Motard (1981) pioneered an evolutionary synthesis approach for separation processes, emphasizing the iterative generation and refinement of potential configurations to find the best-performing sequence. Nadgir and Liu (1983) developed heuristic methods that allowed for a systematic exploration of initial sequences for multicomponent separations. Their work provided a structured approach that helped narrow down the vast design space, making it more manageable for engineers to evaluate different options. Similarly, Malone et al. (1985) proposed analytical criteria for the sequencing of distillation columns, which were aimed at simplifying the decision-making process by identifying key factors that influence the efficiency and cost-effectiveness of different configurations. Wright's early patent on fractionation apparatus also laid the foundation for innovative designs that have influenced many modern distillation technologies, particularly those focused on enhancing energy efficiency (Wright, 1949).

In addition to these heuristic and analytical approaches, more advanced methodologies have been developed to address the need for a systematic and comprehensive generation of distillation sequences. Caballero and Grossmann (2001) proposed a generalized disjunctive programming model that allows for the optimal synthesis of thermally linked distillation columns. This approach, using optimization techniques, systematically generates potential configurations, providing a pathway to find energy-efficient solutions that might not be immediately apparent using traditional methods. The concept of thermodynamically equivalent configurations, also introduced by Caballero and Grossmann, has further expanded the range of feasible solutions by demonstrating that different structural arrangements of a distillation system can have identical thermodynamic efficiency, thereby increasing flexibility in design choices.

Despite these advancements, the search for the most effective distillation configuration remains a challenging task, particularly as the number of components and the need for energy-saving strategies increase. There is still a pressing need for robust methodologies that can systematically generate all feasible conventional designs for separating N components, while also incorporating advanced energy-saving techniques such as thermal coupling. Such methodologies would enable engineers to fully explore the design space and identify the most cost-effective and energy-efficient solutions, ultimately advancing the field of separation processes and contributing to more sustainable industrial practices.

Separation of multicomponent mixtures

For the separation of a mixture containing N components, the general rule is that N-1 distillation columns are required. This principle forms the foundation of conventional multicomponent distillation design, providing a starting point for determining the number of columns necessary to achieve complete separation. Each column is responsible for separating one component at a time, reducing the mixture to pure components step by step.

One of the seminal works in this area is the study conducted by Thompson & King (1972), which focused on the systematic synthesis of separation schemes for multicomponent distillation. Their research was groundbreaking in that it provided a structured methodology to understand and predict the complexity of designing distillation sequences for multicomponent mixtures. Thompson and King explored how different sequences of distillation columns could be arranged to achieve the desired separations, taking into account the numerous possible ways that components could be split at each stage of the process.

Their work culminated in the derivation of a formula to calculate the number of possible simple column configurations (S_n) for a given N-component mixture:

$$S_n = \frac{[2(N-1)]!}{N!(N-1)1} \tag{4.1}$$

This formula is significant because it illustrates the factorial growth in the number of feasible column configurations as the number of components increases. Specifically, as the number of components (N) in the mixture grows, the number of possible sequences that need to be evaluated grows rapidly, making the design process increasingly complex. For example, for a mixture containing four components, Thompson and King demonstrated that there are five distinct configurations of simple columns that can be used to achieve the desired separations.

The methodology proposed by Thompson and King involved systematically generating all possible separation sequences for a given mixture and then evaluating their feasibility and efficiency. By doing so, they were able to identify configurations that were not only practical but also cost-effective in terms of both capital and operational expenses. Their approach highlighted the importance of considering multiple design alternatives rather than settling for a single, potentially suboptimal, configuration.

In addition to providing the mathematical framework for calculating the number of configurations, Thompson and King's work also emphasized the need for flexibility in distillation system design. They recognized that the optimal configuration could depend on various factors, such as feed composition, energy costs, and the desired purity of the products. By developing a systematic way to generate and evaluate different configurations, they paved the way for more sophisticated approaches to distillation design that take into account both technical and economic considerations.

Thompson and King's contributions remain highly influential in the field of chemical engineering, particularly in the context of multicomponent distillation. Their work laid the foundation for subsequent research into the optimization of distillation sequences and the development of more advanced techniques for synthesizing separation processes. Today, their formula and methodological approach are still used as a benchmark for understanding the complexities of multicomponent distillation and for guiding the development of efficient separation schemes.

In the following sections, we will present a methodology proposed by Errico et al. (2009), which focuses on the systematic and ordered generation of distillation sequences, extending beyond conventional configurations to include thermally coupled, thermodynamically equivalent, and intensified structures. The approach proposed by Errico et al. (2009) allows for the comprehensive exploration of both conventional and advanced distillation configurations, providing a robust framework for identifying energy-efficient solutions that address the limitations of traditional column sequences. This systematic approach includes the generation of intensified structures with fewer columns, reducing capital investment and offering the potential for significant energy savings, making this methodology highly relevant for modern, sustainability-driven chemical processes.

Considering a mixture of five components, Fig. 4.1 shows all the alternatives generated.

Thermally coupled alternatives

Thermal coupling is a strategy used in distillation to enhance energy efficiency by reducing the number of reboilers and condensers required in a system. By thermally linking columns, it is possible to create more energy-efficient systems that minimize energy consumption while maintaining separation performance.

Notable contributions to the study of thermally coupled distillation systems include the works of Rong and Kraslawski (2002) and Rong et al. (2003). In their study, Rong et al. (2003) examined partially thermally coupled distillation systems for multicomponent separations, providing insights into the benefits and challenges associated with implementing partial thermal couplings. Their findings demonstrated that thermal couplings could significantly reduce energy requirements and capital costs, particularly for systems involving a large number of components. In their study, Rong and Kraslawski (2002) further

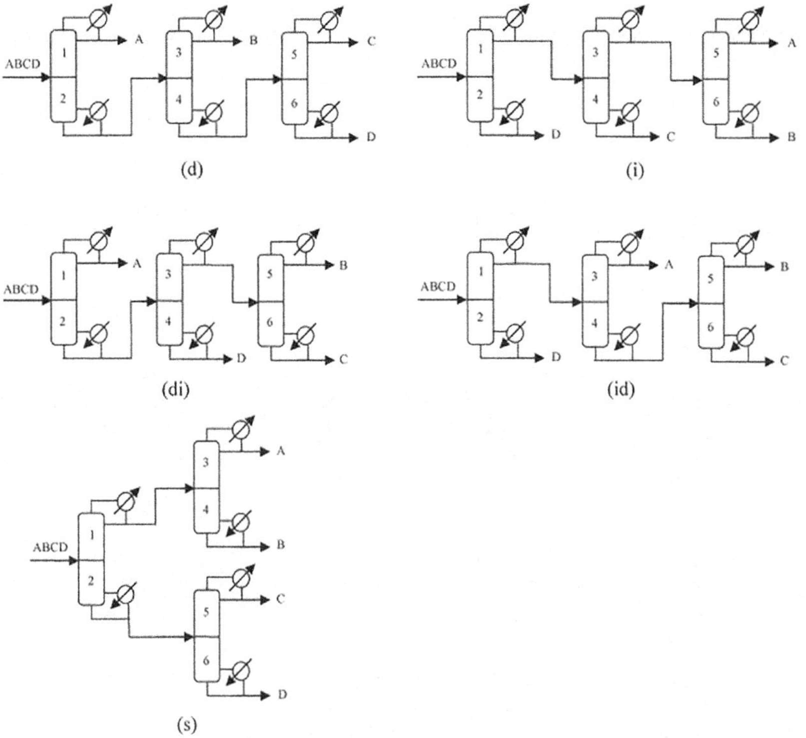

FIGURE 4.1 Conventional distillation alternatives for a four-component mixture, (d) for direct, (i) for indirect, (di) for direct-indirect, (id) for indirect-direct, (s) for symmetrical, respectively.

developed optimal design methodologies for distillation flowsheets with fewer thermal couplings. This work contributed to the understanding of how to balance the complexity and cost of thermal coupling with the potential energy savings. Their research concluded with the derivation of a formula, which is crucial for understanding the number of thermally coupled configurations.

The formula derived for calculating the total number of thermally coupled configurations (P_n) for a given N-component mixture () is as follows:

$$P_n = \frac{[2(N-1)]!}{N!(N-1)1}\left[\sum_{j=1}^{n-3}\frac{(N-2)!}{j!(N-2-j)j!}\right] \tag{4.2}$$

This formula highlights the potential reduction in the number of columns and thermal couplings required to achieve separation, emphasizing the advantages of using thermally coupled configurations to improve energy efficiency and reduce capital costs. The generation of thermally coupled configurations involves substituting one or more condensers and/or reboilers, which are associated with non-product streams, with bidirectional vapor–liquid connections. This transformation results in a subset of thermally coupled sequences, which can offer significant energy savings compared to conventional configurations. The iterative process of modifying simple column configurations to incorporate thermal couplings introduces additional degrees of freedom that can be exploited to enhance the energy efficiency of the overall system. By systematically generating and evaluating thermally coupled configurations, engineers can explore a broader range of design alternatives, ultimately leading to more energy-efficient and cost-effective separation processes.

The findings from Rong and Kraslawski, combined with the systematic generation of thermally coupled configurations, underscore the importance of considering thermal couplings as a viable option for multicomponent separations. The integration of thermal couplings provides a pathway to achieving significant energy savings, particularly in complex systems where the number of components and separation stages is high. This approach serves as a critical reference for understanding the potential of thermal coupling in distillation and provides a methodological framework for generating these configurations in a systematic and efficient manner. Fig. 4.2 displays all the thermally coupled alternatives that can be obtained from the conventional alternatives in Fig. 4.1.

Thermodynamically equivalent configurations

Thermodynamically equivalent configurations are an advanced design concept in distillation aimed at achieving energy efficiency through the rearrangement of column sections, providing flexibility without compromising the overall separation performance. Unlike thermally coupled systems that rely on direct thermal links, thermodynamically equivalent configurations rearrange column sections to redistribute heat and mass flows effectively while maintaining similar energy requirements.

A thermodynamically equivalent design can be obtained by making specific modifications to a thermally coupled structure, particularly by moving a column section associated with a condenser or reboiler that facilitates the common reflux or boil-up between consecutive columns. In this process, either the condenser or reboiler from a submixture can be eliminated, making the corresponding rectifying or stripping section of the subsequent column movable. This rearrangement introduces degrees of freedom that enhance operational flexibility and can contribute to capital cost reduction. Essentially, while thermally coupled columns are connected by shared energy, thermodynamically equivalent configurations achieve similar integration by altering the sequence without necessarily depending on shared heat exchange equipment.

The physical differences between thermally coupled and thermodynamically equivalent configurations involve changes in how vapor and liquid

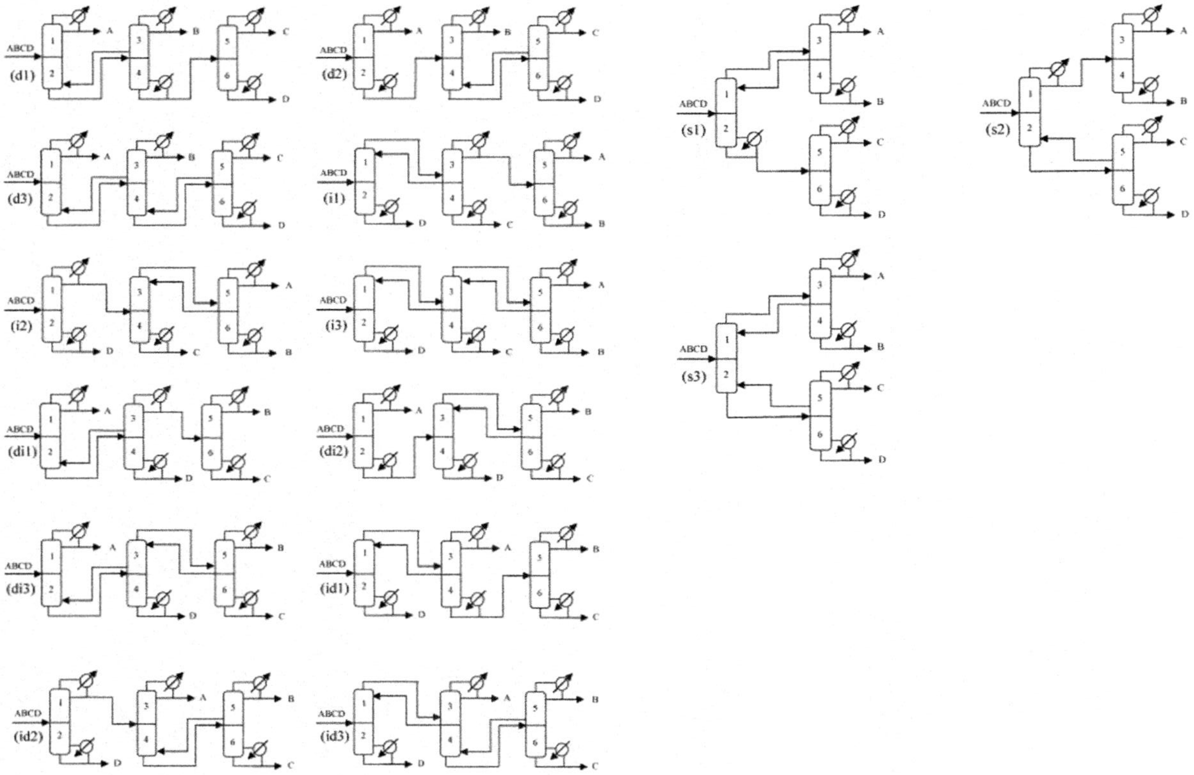

FIGURE 4.2 Thermally coupled distillation configurations for a four-component mixture, (d) for direct, (i) for indirect, (di) for direct–indirect, (id) for indirect direct, and (s) for split scheme.

streams are redistributed. While thermally coupled configurations have direct bidirectional vapor-liquid connections, thermodynamically equivalent ones substitute these with indirect flow redistribution, which can simplify the mechanical aspects of the design and improve system controllability. The removal of some thermal couplings makes these configurations simpler to operate and maintain, with fewer potential points of failure, enhancing reliability.

Thermodynamically equivalent configurations also offer potential advantages over their thermally coupled counterparts in terms of capital cost savings. By eliminating direct thermal connections, the design reduces the number of shared reboilers and condensers, thereby lowering equipment costs and simplifying maintenance. Additionally, improved control characteristics make these configurations more responsive to disturbances, providing an edge in operational environments with fluctuating conditions.

The number of thermodynamically equivalent structures (*TES*) that can be generated from a given thermally coupled sequence is predicted by Eq. (4.3):

$$TES = 2^{NTC} \tag{4.3}$$

where, *NTC* is the number of thermal couplings present in the original sequence. This formula highlights how the number of possible thermodynamically equivalent configurations grows exponentially with the number of thermal couplings. Therefore, systematically generating these configurations allows for the exploration of a broader design space, potentially leading to enhanced operational benefits such as improved control, reduced capital costs, and minimized energy consumption. At the beginning, the subspace of TESs is usually not explored when searching for the best energy-saving structure, as the energy requirement of these structures is the same as that of the thermally coupled configurations from which they are derived. However, once a promising thermally coupled configuration is identified, the corresponding TES should be considered for potential capital cost reduction. Fig. 4.3 presents all the TES generated from the thermally coupled configurations shown in Fig. 4.2. In total, 25 TES were generated.

Intensified alternatives with less of N-1 columns

A distillation structure with less than N-1 columns is referred to as an "intensified design." These designs are part of an effort to reduce the physical and energy requirements of traditional distillation setups. In intensified designs, the number of columns is reduced by eliminating columns that have only one section, typically by substituting condensers or reboilers associated with nonproduct streams with side streams, either in the vapor or liquid phase. This process allows for fewer columns than the conventional minimum of N-1 columns required for separating N components.

The physical changes in an intensified design mainly involve eliminating redundant side columns that have limited functionality. For instance, side

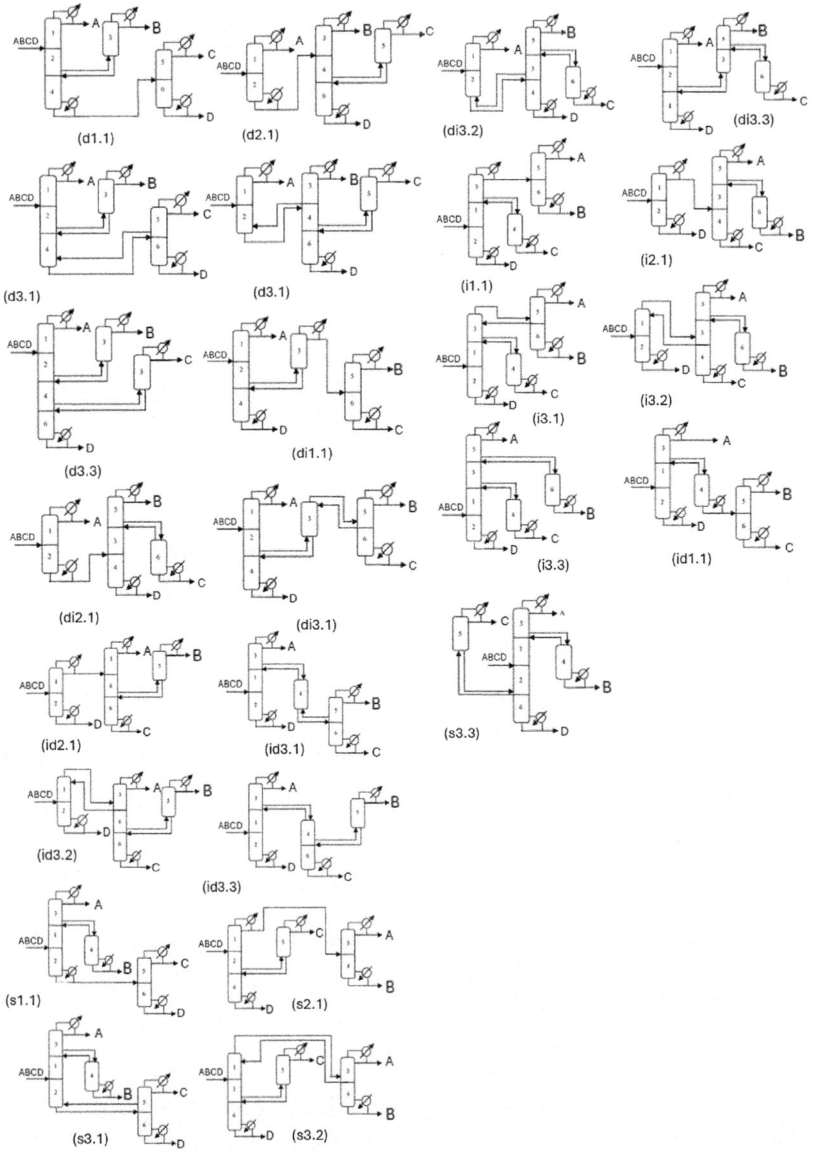

FIGURE 4.3 Thermodynamically equivalent configurations for a four-component mixture, (d) for direct, (i) for indirect, (di) for direct–indirect, (id) for indirect direct, and (s) for split scheme.

sections that contribute minimally to separation can be removed, leading to a reduction in the total number of columns and thermal units. This simplification is accomplished by utilizing side streams in place of previously coupled thermal systems.

There are several potential advantages to these intensified designs. By reducing the number of columns, the capital and operational costs are decreased while maintaining separation efficiency. Due to the simplified structure, fewer control loops are needed, which enhances the overall stability and flexibility of the separation process.

The number of intensified structures that can be generated from a given thermodynamically equivalent sequence (denoted as *NIS*) can be predicted by the following equation:

$$NIS = 2^{NS} - 1 \tag{4.4}$$

where *NS* represents the number of single-section columns in the TES. This formula indicates the possible configurations that maintain efficiency while reducing the complexity of the system. Fig. 4.4 presents all of the systems with less than N-1 columns generated from the TES shown in Fig. 4.3.

To extend the content based on the discussions regarding side streams, a consideration on the phase (vapor or liquid) of the withdrawn side streams is crucial for achieving efficient designs with fewer than N-1 columns. According to the guidelines suggested by Luyben (2013), the phase of the side stream is determined by the concentration distribution of the components involved. If the lightest component is present in small quantities, it is preferable to withdraw the middle product as a liquid stream in the rectifying section (above the feed). This is because the middle component is typically more concentrated in the liquid phase, whereas the lightest component has a higher concentration in the vapor phase. Similarly, in the stripping section (below the feed), the heaviest component tends to be more concentrated in the liquid phase. Therefore, it is recommended to withdraw the middle product from the vapor phase.

In the current approach, the selection of the vapor or liquid phase for the withdrawn side streams is determined by the state of the stream connecting the side column to the main column. If the side stream replaces a rectifying section, it is typically in the vapor phase; conversely, if it replaces a stripping section, it is generally in the liquid phase. This distinction is important for optimizing both energy consumption and process operability.

Case study: Biobutanol purification from a fermentation broth

Biobutanol is an alcohol produced biologically, primarily through the fermentation of sugars by Clostridium species in a process known as acetone-butanol-ethanol (ABE) fermentation (Jones & Woods, 1986). It is an emerging biofuel with significant promise due to its high energy content and compatibility with existing fuel infrastructure. Unlike ethanol, biobutanol can be used in gasoline engines without modifications, which makes it an attractive alternative fuel for reducing greenhouse gas emissions and dependency on

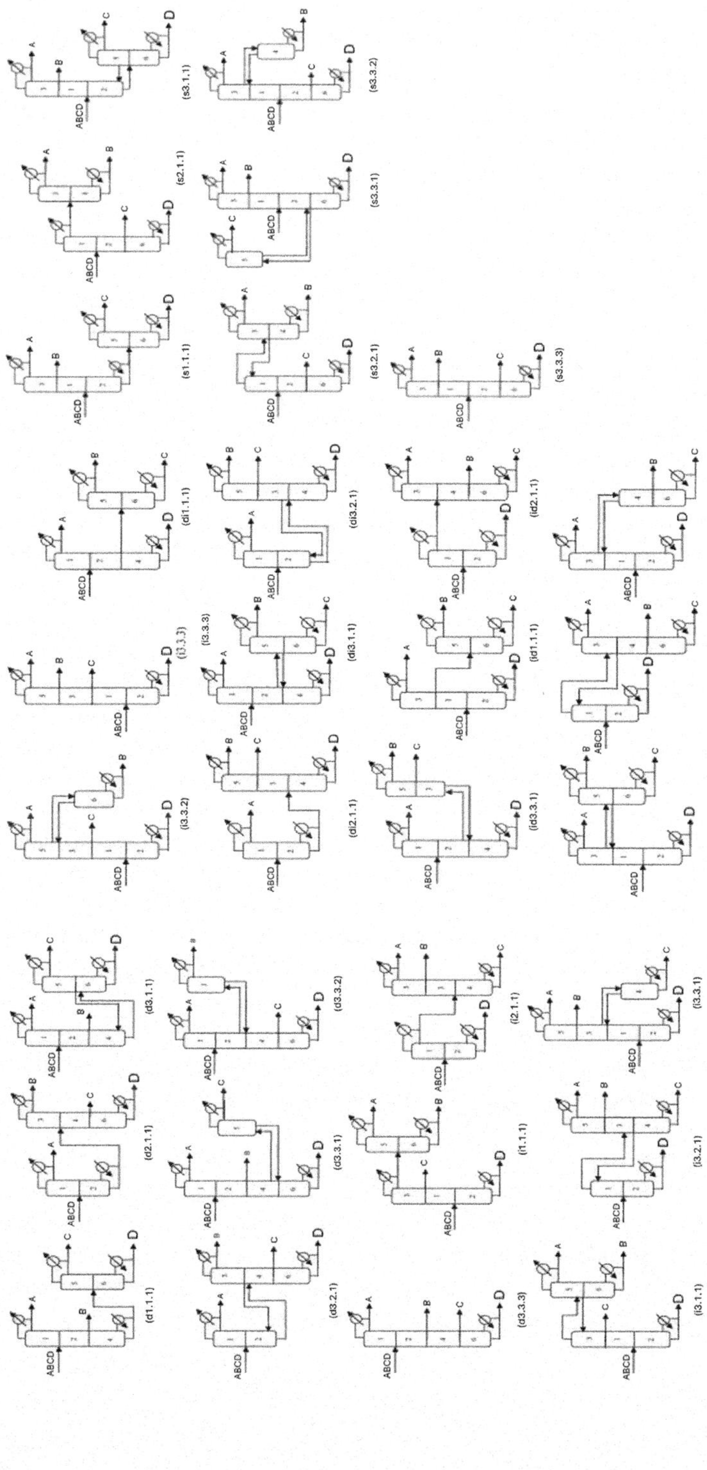

FIGURE 4.4 Intensified alternatives with less than N-1 columns for a four-component separation.

fossil fuels (Dürre, 2007). The importance of biobutanol in today's society stems from its potential role in energy transition toward renewable and sustainable resources, contributing to energy security while reducing environmental impact (Kumar & Gayen, 2011). As the world faces challenges related to climate change and energy demand, biobutanol has emerged as a key player in the biofuel sector, with particular relevance for transportation and other sectors requiring liquid fuels (Ezeji et al., 2007).

The production of biobutanol is typically accomplished through the fermentation of various biomass feedstocks, including sugars derived from agricultural residues and waste products (Ezeji et al., 2004). During the fermentation process, a mixture known as the ABE broth is obtained, which contains acetone, butanol, ethanol, and a significant proportion of water, alongside other minor by-products (Bankar et al., 2013). The composition of the ABE broth is generally characterized by a high-water content, which poses a significant challenge for the subsequent separation and purification of biobutanol. Typically, the butanol concentration in the fermentation broth is relatively low (around 1%–2% by weight), while water makes up the majority of the mixture, making efficient separation crucial to the economic viability of the process (Groot et al., 1992).

Several separation techniques have been explored for the recovery and purification of biobutanol from the fermentation broth. Conventional distillation is one of the most straightforward methods; however, it is highly energy-intensive due to the low concentration of butanol and the presence of a large amount of water (Mariano et al., 2010). Distillation is effective for separating components based on differences in volatility but requires substantial energy input, which makes it less favorable for dilute systems such as ABE fermentation broths. Liquid-liquid extraction (LLE) has been proposed as an alternative, leveraging the use of an organic solvent to selectively extract butanol from the aqueous mixture (Ezeji et al., 2004). The main advantage of LLE is its lower energy requirement compared to distillation, as it bypasses the need to vaporize the water. However, the efficiency of LLE is highly dependent on the choice of solvent, and the subsequent solvent recovery steps can add complexity and cost to the process.

Pervaporation, a membrane-based separation technique, has also been explored for butanol recovery. This method involves the selective permeation of butanol through a membrane, followed by its evaporation on the permeate side. Pervaporation offers the advantage of being energy-efficient and capable of breaking azeotropes, but its limitations include membrane fouling and scalability issues, which can hinder its industrial application. Adsorption, using materials such as activated carbon or zeolites, has also been investigated for biobutanol recovery. Adsorption can be effective for removing butanol from dilute solutions, but the regeneration of adsorbents and the adsorption capacity are key factors that impact the overall feasibility of the process (Xue et al., 2016).

Hybrid separation processes have gained significant attention as they combine the strengths of multiple separation techniques to achieve efficient and cost-effective purification. One such hybrid approach involves the use of a LLE column in combination with distillation columns (Errico et al., 2016). In this configuration, LLE is used to first extract butanol from the aqueous phase of the fermentation broth, significantly reducing the water content before further purification is carried out by distillation. This hybrid method effectively addresses the energy demands of distillation by reducing the volume of water that needs to be vaporized, thereby enhancing overall process efficiency (Kraemer et al., 2011). Additionally, the LLE step provides a concentrated butanol stream, which facilitates subsequent distillation and reduces the operational costs associated with the separation. These hybrid separation strategies are particularly suitable for separating biobutanol from the ABE fermentation effluent. The combination of an LLE column with distillation columns provides an efficient route to recover biobutanol, as it leverages the benefits of both extraction and distillation, allowing for a more energy-efficient process while maintaining high product purity. The use of such hybrid systems is especially relevant for processes dealing with large volumes of dilute solutions, such as those obtained from fermentation, where energy efficiency and economic feasibility are crucial for commercial success (Errico et al., 2017).

In the following section, the methodology previously explained by Errico et al. (2009) will be applied to the ABE effluent to separate and purify biobutanol for use as a fuel. This approach aims to evaluate the potential of intensified configurations in achieving the desired separation performance with reduced energy consumption and capital cost, thereby enhancing the overall feasibility of biobutanol production for fuel applications.

Hybrid alternatives for biobutanol purification

This section focuses on the separation of biobutanol from the ABE effluent using different intensified configurations for efficient purification. The following design alternatives are generated based on the systematic synthesis methodology introduced by Errico et al. (2009), tailored for the biobutanol separation process.

The mixture that will be separated in the case study consists of a fermentation effluent known as the ABE (ABE) mixture. The components of the ABE mixture include butanol, acetone, ethanol, and water. The complete data for the mixture is presented in Table 4.1.

The base separation strategy consists of a hybrid approach, which combines LLE and distillation to achieve the efficient recovery and purification of biobutanol. In the LLE column, hexyl acetate is used as the solvent for extracting butanol from the fermentation broth due to its favorable partition coefficient and immiscibility with water. This base configuration significantly

TABLE 4.1 Composition and characteristics of the mixture to be separated.

Component	Concentration (wt.%)	Remarks
Butanol	0.3018	High boiling point
Acetone	0.1695	Intermediate boiling point
Ethanol	0.0073	Low boiling point
Water	0.5214	Major component

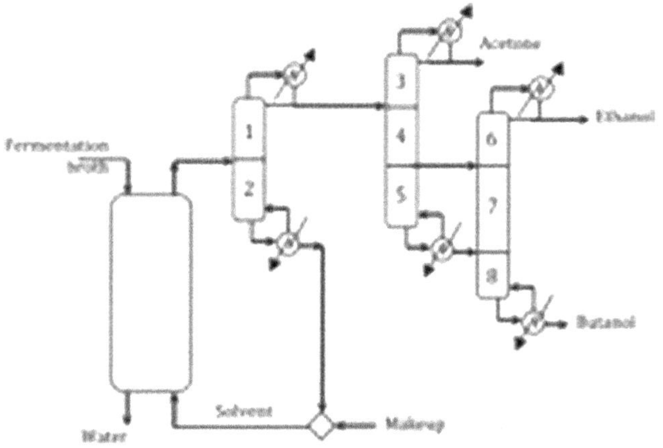

FIGURE 4.5 Hybrid alternative taken as reference configuration.

reduces the water content before subjecting the butanol-rich extract to distillation for further purification. On top of this base configuration, process intensification strategies are applied to explore more energy-efficient alternatives (See Fig. 4.5).

Design strategy and methodology

The synthesis methodology applied to the base hybrid configuration involves three key design strategies, as described below:

1. Alternative Thermally Coupled Configurations:

Starting with the hybrid LLE and distillation structure, thermal couplings are introduced to reduce the number of reboilers and condensers required. These modifications, such as coupling the condenser of the solvent recovery column with the reboiler of another distillation column, help in minimizing energy consumption. Fig. 4.6 illustrates various thermally coupled configurations derived from the base structure.

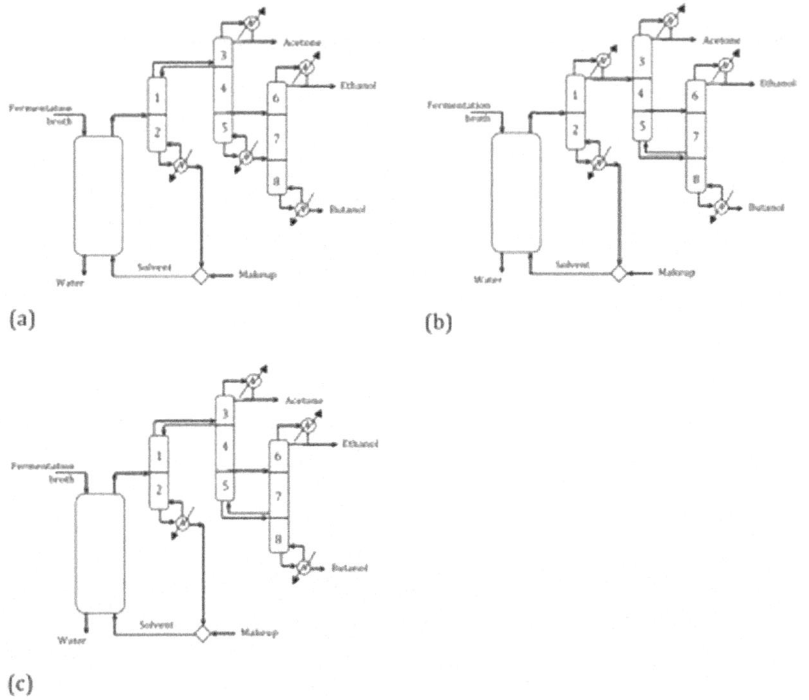

FIGURE 4.6 All thermally coupled alternatives derived from Fig. 4.5.

2. Thermodynamically Equivalent Configurations:

Once the thermally coupled configurations are defined, thermodynamically equivalent configurations are generated by repositioning specific sections of the distillation columns. This allows for maintaining similar thermal efficiency while modifying the physical sequence of operations, often simplifying control and operational requirements. In total, five equivalent alternatives were generated, providing additional degrees of flexibility for optimization. Fig. 4.7 illustrates various thermodynamically equivalent configurations derived from the thermally coupled schemes.

3. Intensified Configurations with Less than N-1 Columns:

Further process intensification is achieved by eliminating redundant sections within the distillation sequence, specifically sections that primarily transport materials rather than separate components. This results in configurations with fewer than N-1 columns, where N represents the number of components in the mixture. Such designs significantly reduce the equipment footprint and associated capital costs while maintaining high separation efficiency. Fig. 4.8 illustrates various intensified configurations derived from the thermodynamically equivalent alternatives.

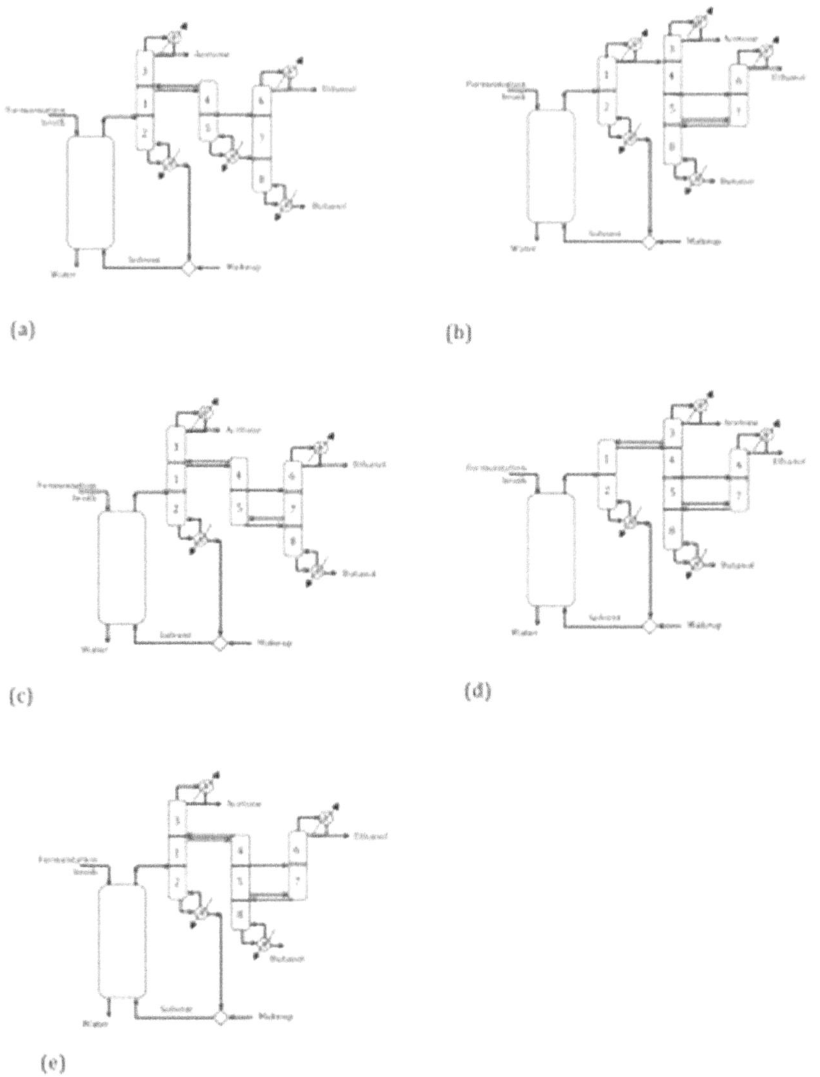

FIGURE 4.7 Thermodynamically equivalent alternatives. Five combinations are possible. (a) It was obtained from the corresponding thermally coupled sequence of Fig. 4.6a moving section 3 above section 1. Following the same procedure, (b) was generated moving section 8 below section 5. Configurations shown in (c)–(e) are obtained from the corresponding sequence in Fig. 4.6c, where the presence of two thermal couplings makes sections 3 and 8 simultaneously movable.

Simulation procedure for the configurations

The newly synthesized configurations were tested through rigorous process simulation using Aspen Plus. The simulations involved defining the feed properties based on the composition of the fermentation broth and

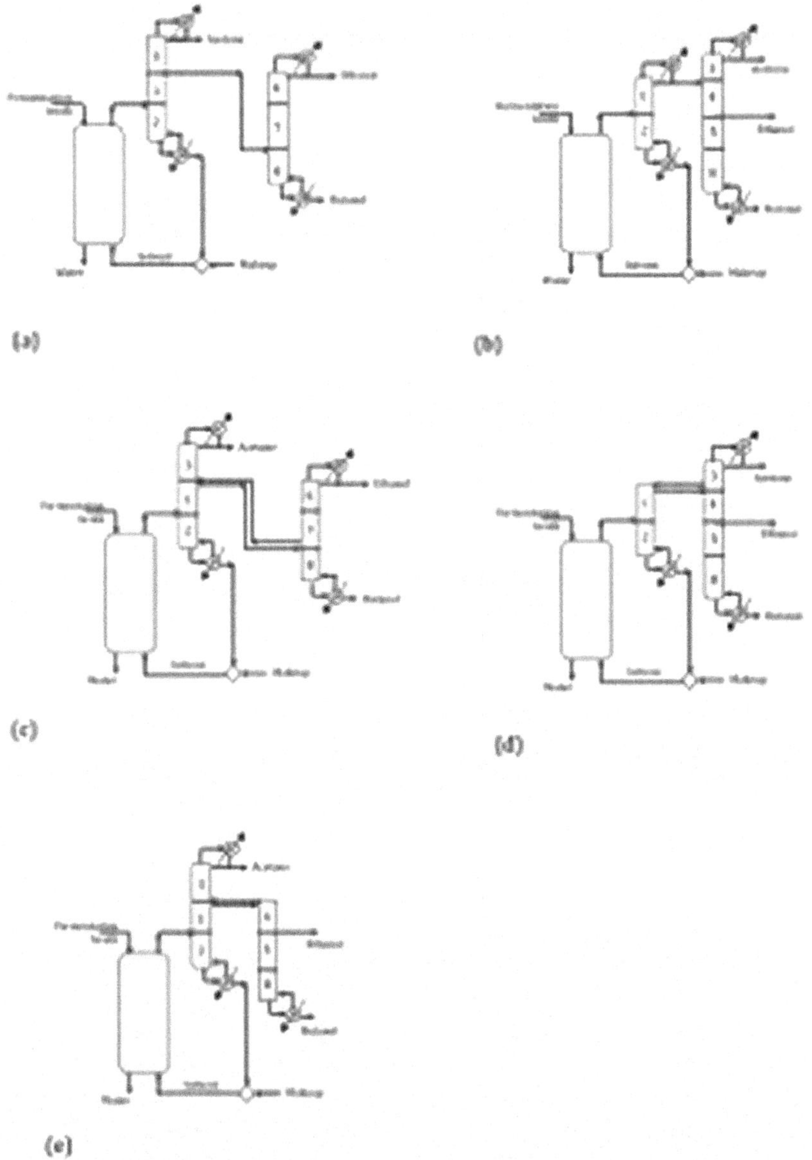

FIGURE 4.8 Intensified configurations derived from Fig. 4.7. For instance, the configuration shown in (a) was obtained from the corresponding sequence of Fig. 4.7a by elimination of sections 4 and 5 and connecting the solvent recovery column to the ethanol/butanol separation column. The same principle has been used for the sequences in part figures (b)–(e).

implementing the configurations in the simulator to evaluate their performance under various operating conditions. The NRTL-HOC thermodynamic model was employed to accurately predict phase equilibria, particularly for the nonideal interactions in the ABE mixture. Hexyl acetate was chosen as the solvent in the LLE column, considering its compatibility with the other components.

The minimum target purity for biobutanol was set at 99.5 wt.%, while acetone and ethanol were also purified to similar levels to ensure that the final product could meet fuel specifications. Tables 4.2 and 4.3 summarize the feed characterization and thermodynamic model used for the simulations.

The economic and environmental performance of the generated configurations was evaluated using the following indices:

1. Economic Objective Function: The total annual cost (*TAC* was used as the primary economic index, which encompasses capital investment and operating expenses for each configuration (Eq. 4.5). The goal was to minimize *TAC* to identify cost-effective designs.

$$TAC = \frac{capital \quad cost}{n} + services \tag{4.5}$$

TABLE 4.2 Design parameters and comparison indexes for the reference configuration.

	Extractor	C1	C2	C3
Number of theoretical stages	5	26	46	20
Reflux ratio	–	0.905	6.034	14.836
Feed stage	1	13	32	5/15
Solvent feed stage	5	–	–	–
Side stream stage	–	–	44	–
Column diameter (m)	0.335	0.322	0.325	0.292
Operative pressure (kPa)	101.353	101.353	101.353	101.353
Distillate flowrate (kg/h)	–	21.687	7.694	0.333
Side stream flowrate (kg/h)	–	–	1.901	–
Solvent flowrate (kg/h)	708.549	–	–	–
Solvent makeup (kg/h)	0.709	–	–	–
Condenser duty (kW)	–	7.284	7.736	1.239
Reboiler duty (kW)	–	65.919	8.428	0.907

TABLE 4.3 Design parameters and comparison indexes for the intensified alternative.

	Extractor	C1	C2
Number of theoretical stages	5	58	20
Reflux ratio	–	27.182	–
Feed stage	1	45	–
Solvent feed stage	5	–	–
Side stream stage	–	–	12
Column diameter (m)	0.335	0.323	0.324
Operative pressure (kPa)	101.353	101.353	101.353
Distillate flowrate (kg/h)	–	7.711	–
Thermal coupling flowrate (kg/h)	–	118.621	–
Side stream flowrate (kg/h)	–	–	0.336
Solvent flowrate (kg/h)	708.289	–	–
Solvent makeup (kg/h)	0.684	–	–
Condenser duty (kW)	–	31.094	0.000
Reboiler duty (kW)	–	65.642	24.517

2. Environmental Objective Function: The environmental impact of each configuration was measured using the Eco-indicator 99, which assesses the potential environmental damage caused by emissions, resource use, and other factors. This indicator aggregates 11 impact categories into three major damage categories: human health, ecosystem quality, and resource depletion.

$$EI99 = \sum_b \sum_d \sum_{k \in K} \delta_d \omega_d \beta_b \alpha_{b,k} \tag{4.6}$$

Here, β_b denotes the total quantity of chemical b released per unit of reference flow due to direct emissions, represents the damage caused by category k per unit of chemical b released into the environment, is the weighting factor for damage in category d, and is the normalization factor for damage in category d. This approach considers the impact of steel used for construction, steam used for heating, and electricity used for pumping.

The objective was to find configurations that offer not only economic but also environmental benefits.

Optimization strategy

To optimize the proposed configurations, a multi-objective optimization method combining Differential Evolution (DE) and Tabu Search (TS) metaheuristics was employed. The DE method involves generating new solutions via mutation and crossover operators, followed by evaluating the fitness of each solution and selecting the best-performing individuals for the next generation. To improve the efficiency of the search process, TS was incorporated to keep track of previously visited solutions, thereby avoiding re-evaluation of the same configurations (Srinivas & Rangaiah, 2007).

The optimization was implemented using a hybrid platform that integrated Microsoft Excel and Aspen Plus via dynamic data exchange. Aspen Plus simulated the process for each configuration, and the results were sent back to Excel for evaluation and subsequent optimization steps. Parameters such as column stages, reflux ratios, and feed stages were iteratively adjusted during the optimization to achieve the best balance between TAC and environmental impact.

In the following section, the detailed results of the optimization, including a comparison of economic and environmental performance for each configuration, are presented, with emphasis on the potential benefits of thermally coupled and intensified sequences for biobutanol purification.

Results

This section presents the optimization results for the separation alternatives generated for biobutanol purification, focusing specifically on the thermally coupled, thermodynamically equivalent, and intensified configurations.

The reference configuration, which is a hybrid combination of a LLE column and subsequent distillation columns, serves as the baseline for comparing other alternatives. The detailed design of the reference configuration, including the energy consumption, is summarized in Table 4.2. Acetone was recovered with a purity of 99.78%, ethanol at 95.95%, and butanol at 99.92%.

The thermally coupled alternatives presented in Fig. 4.2 were analyzed for their potential to further reduce energy requirements. The TAC of these thermally coupled configurations was lower compared to the reference structure, with the best configuration (Fig. 4.2c) showing a TAC reduction of 24.5% and an Eco-indicator 99 reduction of 11.8% compared to the hybrid reference structure. This configuration was derived by removing the condenser from the first column and the reboiler from the third column, resulting in a more energy-efficient and economically favorable design.

The Pareto-optimal solutions for the intensified configurations are illustrated in Fig. 4.9 where the trade-off between economic and environmental objectives is highlighted. Among all the thermally coupled alternatives,

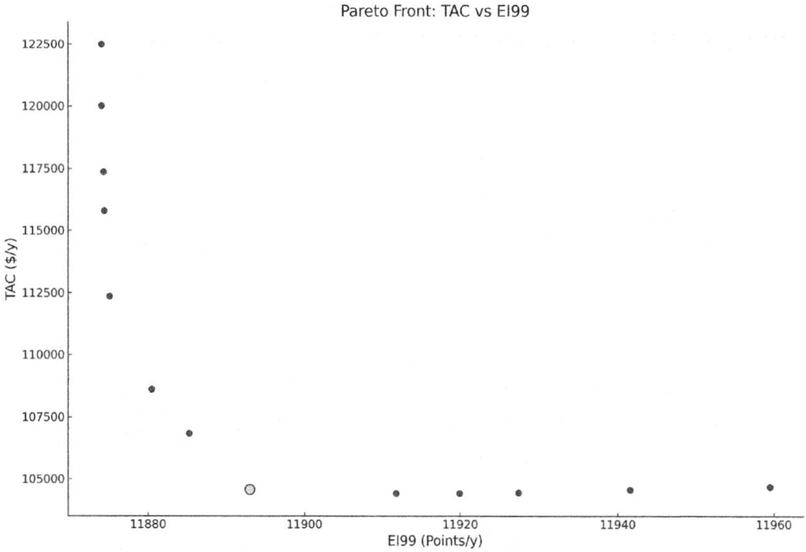

FIGURE 4.9 Pareto front of the intensified alternative.

Fig. 4.2c proved to be the best-performing configuration regarding both TAC and Eco-indicator, which indicates the balanced advantage of this design.

The thermodynamically equivalent configurations derived from the best thermally coupled design, were then evaluated. Although these configurations exhibit similar energy requirements to the thermally coupled configurations from which they were derived, their primary advantage lies in potentially reduced capital costs and simplified operations. Fig. 4.9 presented the best performance among these alternatives, showing the lowest TAC and competitive environmental impact compared to the reference and thermally coupled configurations. The intensified configurations are the final step in the synthesis approach, generated by further simplifying the TES through the elimination of unnecessary column sections. The intensified alternatives (Fig. 4.8c–e) showed promising results, particularly configuration Fig. 4.8e, which achieved the lowest TAC, with a 15% reduction compared to the best thermally coupled option (Fig. 4.2c). However, this intensified configuration exhibited a 25% penalty in the Eco-indicator due to increased reboiler duty, which resulted in higher carbon dioxide emissions and subsequently increased the environmental impact. The intensified structures remain a strong candidate when considering economic optimization, but the environmental trade-offs require careful consideration depending on the specific priorities of the application. Table 4.3 presents the main design parameter for the intensified alternative.

The Pareto-optimal solutions for the intensified configurations were determined, as shown in the newly generated Fig. 4.9. In this figure, each solution represents a balance between minimizing TAC and reducing the

TABLE 4.4 TAC and ecoindicator for selected configurations.

Configuration	TAC ($/year)	Eco-indicator (points/year)
Hybrid (Ref.)	150,277,975	56,975,452
Thermally coupled (Fig. 4.3c)	106,754,390	39,651,947
Intensified (Fig. 4.5e)	105,984,077	40,501,087

environmental impact. The chosen solution, marked by a circle, represents the best compromise between these competing objectives, considering both economic and environmental factors.

The optimization of different biobutanol separation configurations revealed that thermally coupled and thermodynamically equivalent alternatives offer significant benefits over the reference hybrid structure, particularly in reducing TACs (See Table 4.4). Intensified configurations, while economically attractive, presented higher environmental costs, which need to be addressed before implementation. The choice of configuration depends on the specific trade-offs between economic savings and environmental performance that are acceptable for the particular application.

As conclusions for this case study, it is possible to note that biobutanol, produced via fermentation, offers a sustainable alternative to fossil fuels, and optimizing its separation process can lead to significant benefits. The advanced configurations examined—thermally coupled, thermodynamically equivalent, and intensified—demonstrated substantial improvements in energy efficiency and cost. Compared to the reference hybrid setup, the thermally coupled configuration reduced TAC by 24.5% and decreased environmental impact by 11.8%. Intensified configurations achieved even greater TAC reductions (up to 43%), although some came with increased environmental penalties. These results highlight that implementing thermally coupled and intensified designs can make biobutanol production more cost-effective and sustainable, promoting its feasibility as an alternative fuel source.

Conclusions

This chapter has demonstrated that optimizing distillation systems using advanced configurations such as thermal coupling, thermodynamically equivalent configurations, and intensified designs can significantly enhance both energy efficiency and economic viability. This robust methodology was applied to biobutanol separation as case study. Regarding biobutanol separation from a fermentation broth, the reference hybrid configuration, which combined LLE and distillation, served as a baseline for evaluating the performance of these alternative configurations. Thermally coupled configurations emerged as the most promising, reducing the TAC by 24.5% and decreasing the environmental impact by 11.8% compared to the baseline. Thermodynamically equivalent structures provided additional flexibility with potentially lower capital costs and simplified operations, while maintaining similar energy requirements.

Intensified configurations, although achieving the lowest TAC, posed environmental trade-offs that need careful evaluation, particularly with increased reboiler duties contributing to higher carbon emissions. The case study on biobutanol purification underscored the practical benefits and challenges of applying these advanced techniques, offering valuable insights into the trade-offs between cost savings and environmental sustainability.

In conclusion, the advanced synthesis methodologies discussed in this chapter are crucial for addressing the complexities of multicomponent separation, especially in light of growing demands for sustainable and energy-efficient processes. The implementation of these optimized distillation designs can substantially improve the feasibility of biofuels like biobutanol, advancing the pursuit of more sustainable energy solutions. Future work should focus on the experimental validation of these configurations and further exploration of intensified designs that balance economic and environmental objectives effectively.

References

Bankar, S. B., Survase, S. A., Ojamo, H., & Granström, T. (2013). Biobutanol: The outlook of an academic and industrialist. *Royal Society of Chemistry, Finland RSC Advances, 3*(47), 24734–24757. https://doi.org/10.1039/c3ra43011a

Caballero, J. A., & Grossmann, I. E. (2001). Generalized disjunctive programming model for the optimal synthesis of thermally linked distillation columns. *Industrial and Engineering Chemistry Research, 40*(10), 2260–2274.

Dürre, P. (2007). Biobutanol: an attractive biofuel. *Biotechnology Journal: Healthcare Nutrition Technology, 2*(12), 1525–1534.

Errico, M., Rong, B.-G., Tola, G., & Turunen, I. (2009). A method for systematic synthesis of multicomponent distillation systems with less than N-1 columns. *Chemical Engineering and Processing: Process Intensification, 48*(4), 907–920. https://doi.org/10.1016/j.cep.2008.12.005

Errico, M., Sanchez-Ramirez, E., Quiroz-Ramìrez, J. J., Segovia-Hernandez, J. G., & Rong, B. G. (2016). Synthesis and design of new hybrid configurations for biobutanol purification. *Computers and Chemical Engineering, 84*, 482–492. https://doi.org/10.1016/j.compchemeng.2015.10.009

Errico, M., Sanchez-Ramirez, E., Quiroz-Ramìrez, J. J., Rong, B. G., & Segovia-Hernandez, J. G. (2017). Multiobjective optimal acetone-butanol-ethanol separation systems using liquid-liquid extraction-assisted divided wall columns. *Industrial and Engineering Chemistry Research, 56* (40), 11575–11583.

Ezeji, T. C., Qureshi, N., & Blaschek, H. P. (2004). Butanol fermentation research: Upstream and downstream manipulations. *Chemical Record, 4*(5), 305–314. https://doi.org/10.1002/tcr.20023

Ezeji, T. C., Qureshi, N., & Blaschek, H. P. (2007). Bioproduction of butanol from biomass: From genes to bioreactors. *Current Opinion in Biotechnology, 18*(3), 220–227. https://doi.org/10.1016/j.copbio.2007.04.002

Groot, W. J., van der Lans, R. G. J. M., & Luyben, K. C. A. M. (1992). Technologies for butanol recovery integrated with fermentations. *Process Biochemistry, 27*(2), 61–75. https://doi.org/10.1016/0032-9592(92)80012-R

Jones, D. T., & Woods, D. R. (1986). Acetone-butanol fermentation revisited. *Microbiological Reviews, 50*(4), 484–524. https://doi.org/10.1128/mr.50.4.484-524.1986

Kraemer, K., Harwardt, A., Bronneberg, R., & Marquardt, W. (2011). Separation of butanol from acetone–butanol–ethanol fermentation by a hybrid extraction–distillation process. *Computers and Chemical Engineering, 35*(5), 949–963. https://doi.org/10.1016/j.compchemeng.2011.01.028

Kumar, M., & Gayen, K. (2011). Developments in biobutanol production: New insights. *Applied Energy, 88*(6), 1999–2012. https://doi.org/10.1016/j.apenergy.2010.12.055

Luyben, W. L. (2013). *Distillation design and control using Aspen™ simulation*. Wiley. https://doi.org/10.1002/9781118510193

Malone, M. F., Glinos, K., Marquez, F. E., & Douglas, J. M. (1985). Simple, analytical criteria for the sequencing of distillation columns. *AIChE Journal, 31*(4), 683–689. https://doi.org/10.1002/aic.690310419

Mariano, A. P., Costa, C. B. B., de Angelis, D.d. F., Maugeri Filho, F., Atala, D. I. P., Wolf Maciel, M. R., & Maciel Filho, R. (2010). Optimisation of a continuous flash fermentation for butanol production using the response surface methodology. *Chemical Engineering Research and Design, 88*(5–6), 562–571. https://doi.org/10.1016/j.cherd.2009.11.002

Nadgir, V. M., & Liu, Y. A. (1983). Studies in chemical process design and synthesis: Part V: A simple heuristic method for systematic synthesis of initial sequences for multicomponent separations. *AIChE Journal, 29*(6), 926–934. https://doi.org/10.1002/aic.690290609

Nath, R., & Motard, R. L. (1981). Evolutionary synthesis of separation processes. *AIChE Journal, 27*(4), 578–587. https://doi.org/10.1002/aic.690270407

Rong, B. G., & Kraslawski, A. (2002). Optimal design of distillation flowsheets with a lower number of thermal couplings for multicomponent separations. *Industrial & Engineering Chemistry Research, 41*(23), 5716–5726.

Rong, B. G., Kraslawski, A., & Turunen, I. (2003). Synthesis of functionally distinct thermally coupled configurations for quaternary distillations. *Industrial & Engineering Chemistry Research, 42*(6), 1204–1214.

Seader, J. D., Henley, E. J., & Keith-Roper, D. (2011). *Separation process principles* (3rd ed.). John Wiley & Sons, Inc.

Srinivas, M., & Rangaiah, G. P. (2007). Differential evolution with Tabu list for solving nonlinear and mixed-integer nonlinear programming problems. *Industrial and Engineering Chemistry Research, 46*(22), 7126–7135. https://doi.org/10.1021/ie070007q

Thompson, R. W., & King, C. J. (1972). Systematic synthesis of separation schemes. *AIChE Journal, 18*(5), 941–948. https://doi.org/10.1002/aic.690180510

Underwood, A. J. V. (1949). Fractional distillation of multicomponent mixtures. *Industrial and Engineering Chemistry, 41*(12), 2844–2847. https://doi.org/10.1021/ie50480a044

Wright, R. O. (1949). *Fractionation apparatus*. United States Patent Office (pp. 4–6).

Xue, C., Liu, F., Xu, M., Tang, I. C., Zhao, J., Bai, F., & Yang, S. T. (2016). *Butanol production in acetone-butanol-ethanol fermentation with in situ product recovery by adsorption* (Vol 219, pp. 158–168). Elsevier Ltd, China Bioresource Technology. https://doi.org/10.1016/j.biortech.2016.07.111

Chapter 5

Applications in (bio) thermal coupling processes within a sustainability framework

Use of thermal couplings in productive processes of compounds with non-renewable raw materials

In recent decades, the production landscape in the chemical industry has undergone a significant transformation. Initially, the primary focus was on achieving production targets with minimal regard for energy consumption and production costs. The main objective was to maximize output, often neglecting the environmental and economic repercussions of high energy use. This mindset led to the establishment of processes that, while efficient in terms of production volume, were often highly inefficient in terms of energy consumption and had significant environmental footprints (Smith, 2016).

However, increasing political and social pressures have led to a paradigm shift in the chemical industry. Today, the focus has expanded beyond mere production metrics to encompass sustainability, energy efficiency, and reduced environmental impact. Governments, regulatory agencies, and society at large are demanding that chemical processes become more sustainable, use less energy, and produce fewer emissions. This shift has introduced new challenges and opportunities for the industry, compelling it to reevaluate established processes and adopt new technologies that align with these sustainability goals (Anastas & Warner, 2000).

The importance of certain chemicals in this context cannot be overstated. Many of these chemicals serve as platform compounds for the synthesis of other high-value-added products. Platform chemicals are essential intermediates in the production of a wide variety of industrial products, from polymers to specialty chemicals. These platform chemicals form the backbone of modern industrial chemistry, enabling the manufacture of countless everyday products, including plastics, resins, adhesives, and pharmaceuticals. However, the challenge remains that many of these platform chemicals are still produced from nonrenewable raw materials due to the absence of viable

Thermally Coupled Distillation Columns. https://doi.org/10.1016/B978-0-443-33114-5.00012-0

synthetic routes from renewable feedstocks. Compounds such as diphenyl carbonate, nylon, and various hydrocarbons are typical examples of substances that are primarily produced from fossil-based sources because of the lack of efficient renewable alternatives (Bozell & Petersen, 2010).

Diphenyl carbonate, for example, is a crucial precursor for the production of polycarbonates, which are used in the manufacturing of optical discs, automotive parts, and consumer electronics. The traditional production of diphenyl carbonate involves the use of phosgene, a highly toxic compound, and requires significant energy input, making it an area of concern from both environmental and safety perspectives. Similarly, the production of nylon, a widely used polymer in textiles, automotive, and industrial applications, relies heavily on nonrenewable feedstocks like adipic acid and hexamethylenediamine, both derived from petrochemical sources. The separation of hydrocarbons, essential in the refining of crude oil and in petrochemical processes, is another area where conventional processes are energy-intensive and contribute significantly to greenhouse gas emissions (Arvidsson et al., 2018).

Even though these compounds are derived from nonrenewable resources, the processes used to produce them must now meet stringent energy-saving and sustainability requirements. This shift is largely driven by the need to align chemical production with global sustainability goals, such as those outlined by the United Nations Sustainable Development Goals (SDGs), particularly Goal 12 (Responsible Consumption and Production) and Goal 13 (Climate Action). Consequently, the chemical industry has been under pressure to develop and implement innovative approaches to reduce energy consumption, improve process efficiency, and minimize waste, even for processes that inherently depend on nonrenewable feedstocks (Department of Economic and Social Affairs, 2007).

In this regard, thermal couplings play an essential role in enhancing the sustainability of processes that utilize nonrenewable raw materials. Thermal coupling is a process intensification technique that involves the integration of heat between different stages or units of a process, allowing for substantial energy savings. By effectively managing heat flows, thermal couplings reduce the need for external heating or cooling, thereby decreasing the overall energy consumption of the process. This capability is particularly valuable in processes involving nonrenewable materials, where the focus is on reducing the environmental impact while maintaining the efficiency and output quality (Boodhoo & Harvey, 2013).

The use of thermal couplings has proven to be highly effective in reducing energy requirements in a range of chemical processes. For instance, the production of diphenyl carbonate can benefit from thermal coupling by utilizing excess heat generated in exothermic reactions to drive other process units, thus lowering overall energy consumption. Similarly, in the production of nylon, thermal couplings can be used to optimize the heat integration between polymerization and distillation stages, reducing the need for external heating and improving the process's energy profile. In the refining sector, the separation of complex hydrocarbons is another area where thermal coupling can significantly enhance energy efficiency. By integrating the heat requirements of different distillation columns, refineries can achieve substantial energy savings, contributing to lower operating costs and reduced emissions (Smith, 2016).

Despite being derived from nonrenewable feedstocks, the adoption of thermal couplings in these processes allows the industry to meet modern energy-saving and sustainability demands. The chemical industry, thus, finds itself in a position where it must balance the continued use of nonrenewable resources with the imperative to reduce energy consumption and emissions. Thermal couplings, by virtue of their ability to facilitate heat integration and reduce energy requirements, are a key component of this balancing act. They not only contribute to making existing processes more energy-efficient but also support the industry's broader transition toward more sustainable practices.

Thus, the evolution of production in the chemical industry from a focus on volume and cost to one of sustainability and energy efficiency has necessitated the adoption of new technologies, such as thermal couplings. These technologies are especially critical for processes that rely on nonrenewable feedstocks, helping them to meet modern energy-saving and sustainability demands. The following sections will explore specific case studies—such as the separation of complex hydrocarbons, the production of diphenyl carbonate, and the treatment of effluents from the nylon industry—demonstrating how thermal couplings contribute to optimizing these processes.

Case study: Production of diphenyl carbonate

Diphenyl carbonate (DPC) is an important organic compound used primarily as an intermediate in the production of polycarbonate plastics. Polycarbonates are high-performance materials used in a wide range of applications, including optical discs, automotive components, and electronic devices. The production of DPC is critical to meet the growing demand for polycarbonates, which are known for their excellent transparency, impact resistance, and thermal stability (Antonakou & Achilias, 2013). Given its significance, the development of sustainable and efficient production processes for DPC has become a priority in the chemical industry.

A promising alternative for the production of diphenyl carbonate is the use of a "green" route that employs dimethyl carbonate (DMC) as a starting material. This method is considered environmentally friendly because DMC can be produced from carbon dioxide, which helps in reducing greenhouse gas emissions (Aresta et al., 2014). The transesterification reaction between DMC and phenol yields DPC and methanol as a by-product. This route is often referred to as the "green route" for DPC production due to its potential to minimize the environmental impact associated with traditional methods.

However, despite the environmental benefits, the green route for DPC production faces several challenges that need to be addressed to make it commercially viable. One of the major challenges is the presence of azeotropes in the reaction mixture, particularly involving methanol and DMC, which complicates the separation and purification processes (Sundmacher & Kienle, 2003). Azeotropic mixtures are difficult to separate through conventional distillation, often requiring additional energy-intensive separation techniques or the use of entrainers, which may introduce additional costs and complexity. Moreover, maintaining high selectivity toward DPC is crucial, as side reactions can lead to the formation of undesired by-products that further complicate the purification process and reduce overall yield (Kiss et al., 2019). To overcome these challenges, advanced process configurations are necessary to enhance both the energy efficiency and the selectivity of the reaction. This section will focus on various strategies employed to optimize the production of DPC, with a particular emphasis on reducing energy consumption and improving the sustainability of the process through the use of reactive distillation and thermal coupling techniques.

Reactive distillation as an alternative for DPC production

Reactive distillation (RD) is an innovative process intensification technology that combines chemical reaction and distillation into a single unit. This integration allows for significant process simplification, reduced capital costs, and enhanced energy efficiency (Malone & Doherty, 2000). In the context of DPC production, RD offers the advantage of simultaneously carrying out the transesterification reaction between DMC and phenol while separating the products, thus shifting the equilibrium toward the formation of DPC and improving overall conversion.

In the RD process for DPC production, the key raw materials are phenol and dimethyl carbonate. The process involves two main columns: the reactive distillation column, where the reaction takes place, and a subsequent separation column that serves to purify the DPC and recycle unreacted materials. In the reactive distillation column, the reaction and the separation occur simultaneously, which helps in removing the methanol formed as a by-product from the reaction zone, thereby driving the reaction toward greater conversion. This continuous removal of products enhances reaction efficiency and reduces the need for excess reactants, contributing to improved selectivity and reduced waste generation. The process scheme for this process was previously presented by Contreras-Zarazúa et al. (2017) according to Fig. 5.1.

Despite the inherent advantages of reactive distillation, there is still significant potential for energy optimization. Reactive distillation columns typically require substantial heating and cooling inputs to maintain the desired temperature profiles for both reaction and separation. Therefore, additional

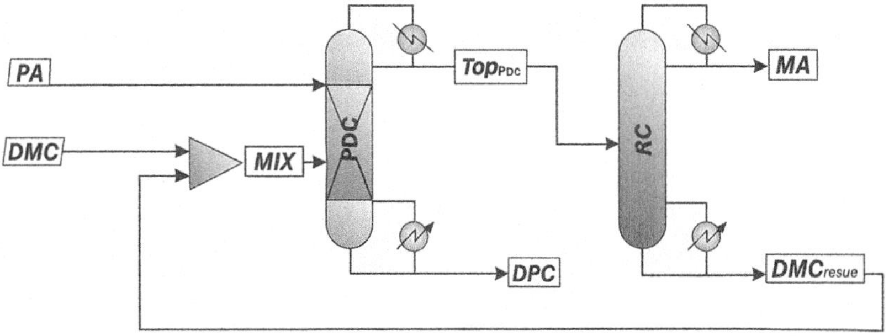

FIGURE 5.1 Conventional reactive distillation scheme (CRD).

strategies, such as thermal coupling, can be applied to further enhance the energy efficiency of the process.

Thermal coupling strategies in reactive distillation

The work of Contreras-Zarazúa et al. (2017) provides a foundational analysis of the production of DPC using a reactive distillation column followed by a separation column, referred to as the base configuration. This base configuration serves as a benchmark for evaluating the performance of alternative energy-saving configurations. Recognizing the potential for energy savings, they proposed and evaluated several advanced configurations, including thermally coupled reactive distillation (TCRD) and reactive distillation with thermally coupled and heat integration (RD-TCHI).

Thermally coupled reactive distillation (TCRD)

The thermally coupled reactive distillation (TCRD) configuration involves integrating a thermal coupling between the reactive distillation column and an auxiliary distillation column. This configuration enables the direct transfer of heat between different sections of the process, thereby minimizing the need for external heating or cooling utilities. The thermal coupling facilitates the use of excess heat from exothermic sections of the process to provide the necessary energy for endothermic sections, thereby enhancing overall energy efficiency.

In the TCRD configuration, the reboiler duty of the main reactive distillation column is partially replaced by heat provided through thermal coupling with another distillation unit. This not only reduces the energy demand but also helps in maintaining the temperature profile within the reactive zone, which is critical for ensuring high selectivity and conversion of DMC to DPC (Contreras-Zarazúa et al., 2017). By reducing the reliance on external heating sources, the TCRD configuration can achieve substantial energy savings, contributing to a lower Total Annual Cost (TAC) for the process (see Fig. 5.2).

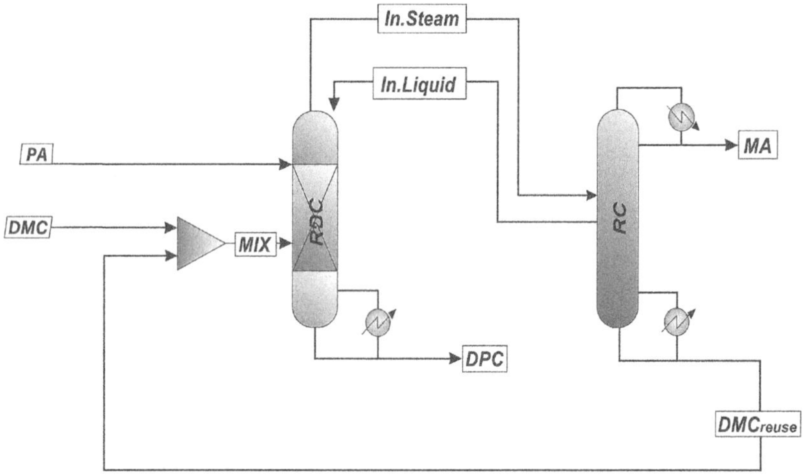

FIGURE 5.2 Thermally coupled reactive distillation scheme (TCRD).

Reactive distillation with thermally coupled and heat integration (RD-TCHI)

Building on the TCRD configuration, the reactive distillation with thermally coupled and heat integration (RD-TCHI) configuration takes energy optimization a step further by incorporating additional heat integration across multiple units within the process. The RD-TCHI configuration not only utilizes thermal coupling between the reactive and separation columns but also integrates heat exchangers that allow for the recovery and reuse of heat from various parts of the process. This comprehensive approach to heat integration ensures that the energy available within the system is utilized to the fullest extent, reducing the need for external utilities even further (see Fig. 5.3).

The RD-TCHI configuration is designed to maximize the use of internal energy sources, such as the heat generated by the exothermic transesterification reaction, to provide the energy required for distillation and separation. This approach significantly reduces both the operating costs and the environmental impact of the process by decreasing fuel consumption and associated greenhouse gas emissions. The enhanced heat integration also contributes to improved process stability, as it allows for better control of temperature gradients throughout the system.

Modeling and performance evaluation

The configurations proposed by Contreras-Zarazúa et al. (2017) were modeled using Aspen Plus, a widely used process simulation software that provides detailed insights into the thermodynamic behavior and performance of chemical processes. The Peng-Robinson equation of state was employed as the thermodynamic model to accurately predict the vapor–liquid equilibria of the components involved, particularly in the presence of azeotropes and complex

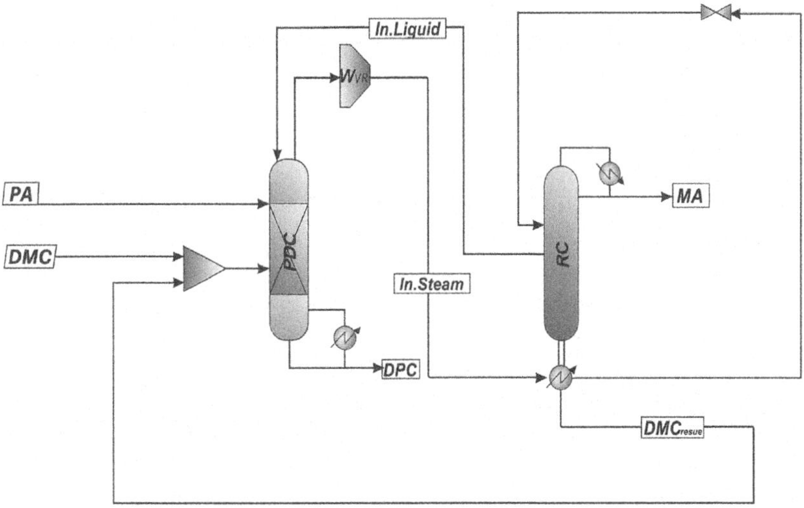

FIGURE 5.3 Reactive distillation with thermally coupled and heat integration scheme (THRD).

interactions between methanol, phenol, DMC, and DPC. To assess the performance of each configuration, a supplementary multi-objective optimization was carried out using differential evolution with a tabu list (DETL). This method enabled a comprehensive exploration of the solution space, facilitating the identification of optimal trade-offs between economic and operational goals. As a result, each process configuration could be evaluated in a more robust and balanced manner.

To evaluate the performance of each configuration, several key indicators were considered:

Total Annual Cost (TAC): This indicator provides a comprehensive measure of the economic viability of the process, taking into account both capital and operational expenditures. TAC is crucial for determining the cost-effectiveness of implementing advanced configurations such as TCRD and RD-TCHI.

Condition Number (CN): The condition number is used to assess the dynamic controllability of the process. A lower CN indicates a system that is easier to control, with fewer fluctuations and a more stable operation, which is essential for maintaining product quality and minimizing downtime.

Results and discussion

The results from the simulation and evaluation studies indicated that both the TCRD and RD-TCHI configurations offered significant improvements over the base configuration. Specifically, the TCRD configuration achieved a reduction in TAC of approximately 15% compared to the base case, primarily due to the reduction in energy consumption achieved through thermal coupling. The RD-TCHI configuration demonstrated an even greater reduction in TAC, up to

25%, as a result of the extensive heat integration that minimized the need for external heating and cooling utilities.

The role of thermal couplings was central to achieving these improvements. By effectively redistributing the heat generated within the process, the TCRD configuration managed to lower the energy requirements for the reactive distillation column. The thermal coupling enabled the utilization of excess heat from the exothermic transesterification reaction, thereby reducing the reboiler duty by approximately 20% compared to the base configuration. This reduction in reboiler duty not only minimized energy costs but also reduced the overall carbon footprint by decreasing fuel consumption.

In the RD-TCHI configuration, the combination of thermal coupling with additional heat integration led to even greater energy savings. By incorporating heat exchangers that recovered and reused heat from various sections of the process, the RD-TCHI configuration reduced utility consumption by up to 30%. This level of energy efficiency was made possible by the strategic use of thermal couplings, which allowed for the effective management of heat flows throughout the system, ensuring that the energy generated by the process was not wasted but instead redirected to where it was needed most. This comprehensive heat integration not only reduced TAC but also contributed to a more sustainable process by lowering greenhouse gas emissions.

Moreover, the condition number analysis highlighted the positive impact of thermal couplings on process controllability. The base configuration exhibited the highest condition number, indicating potential challenges in maintaining stable operation under varying feed conditions or disturbances. The TCRD configuration showed a marked improvement, with a reduction in the condition number by approximately 10% compared to the base case, reflecting enhanced stability due to the effective thermal coupling. The RD-TCHI configuration further reduced the condition number by an additional 15%, demonstrating superior controllability. The role of thermal couplings in achieving these improvements was crucial, as they facilitated more uniform temperature distribution and reduced thermal gradients, which are key factors in maintaining process stability. The enhanced stability observed in the RD-TCHI configuration translated into fewer operational disturbances and a more consistent product quality. The uniform temperature profile across the reactive and separation zones ensured that the reaction proceeded with high selectivity, reducing the formation of by-products and improving the overall yield of DPC. This not only enhanced the economic performance of the process by maximizing the production of the desired product but also contributed to environmental sustainability by minimizing waste generation. Table 5.1 shows the general data for all the alternatives evaluated, performance indexes, and design parameters.

Thus, thermal couplings played a pivotal role in the optimization of the DPC production process. By enabling effective heat management and reducing reliance on external utilities, thermal couplings significantly improved both the

TABLE 5.1 Design parameters and performance indexes for all the process schemes.

Design variable/parameter	CRD	TCRD	THRD
Number of stages, RDC	78	85	80
Number of stages, RC	26	22	23
Feed stage of PA	7	10	9
Feed stage of DMC	71	80	74
Feed stage, RC	11	13	8
Diameter of RDC (m)	1.142	0.990	1.073
Diameter of RC (m)	1.426	0.740	1.021
Reflux ratio, RDC	2.594	–	–
Reflux ratio, RC	2.383	8.3	9.694
Heat duty of RDC (kW)	819	607	514
Heat duty of RC (kW)	161	86	819
Top pressure RDC (atm)	2.383	2.950	2.035
Tray holdup (l)	346.310	437.210	213.231
Reactive stages, RDC	7–74	10–81	4–74
Interlinking flow (kmol/h)	–	36.050	–
Energy of compressor (kW)	–	–	30
Discharge pressure of compressor (atm)	–	–	3
Purity of DPC (mol fraction)	0.995	0.996	0.996
Purity of MA (mol fraction)	0.996	0.997	0.996
Energy per ton of DPC produced (GJ/TonDPC)	3.410	2.890	1.670
Total energy (kW)	980	693	544
Condition number	1764.49	2640.16	1541.90
Total annual cost ($/yr)	1,854,324	906,241	1,734,136

economic and environmental performance of the process. The TCRD and RD-TCHI configurations demonstrated that integrating thermal coupling and heat integration techniques can lead to substantial energy savings, lower costs, and enhanced process controllability. These results underscore the importance of adopting process intensification strategies, such as thermal couplings, to develop sustainable chemical production processes that meet modern energy-saving and sustainability requirements.

Case study: Separation of effluents from the nylon industry

The consumption of chemical products is closely linked to population growth and economic expansion. As economies develop and populations grow, the demand for chemicals in everyday applications—ranging from pharmaceuticals to polymers and industrial solvents—has increased significantly. This growth is directly tied to the generation of chemical waste, which poses substantial risks to environmental health and safety. The continuous rise in chemical production has resulted in the accumulation of hazardous wastes, often containing compounds that are toxic to both ecosystems and human health. This environmental challenge has necessitated the exploration of sustainable solutions to mitigate waste and minimize adverse impacts (Zaidi et al., 1995).

The concept of a circular economy has emerged as a key strategy for addressing these challenges. The main idea behind the circular economy is to eliminate waste by recovering and reintegrating by-products into the production cycle, thus closing the loop and minimizing resource wastage. In the chemical industry, this approach emphasizes converting waste streams into valuable products while improving energy efficiency and sustainability. One effective method to achieve these objectives in separation processes is through the application of energy-saving strategies, such as thermal couplings. Thermal couplings facilitate the efficient exchange of heat between different stages of a process, reducing the need for external utilities, minimizing energy consumption, and improving the sustainability profile of the process. In the context of effluent recovery from the nylon industry, the use of thermal coupling can yield multiple benefits. By recovering a hazardous effluent and converting it into a valuable product, the process aligns with the principles of energy savings, sustainability, and the circular economy. This approach not only provides a method for managing waste streams but also contributes to economic sustainability by producing high-value-added compounds that are in high demand in the chemical market (Zhou et al., 2018).

The nylon industry is a major contributor to global polymer production, with millions of tons of nylon 6 and nylon 66 produced annually. As a result, a substantial amount of waste is generated, often containing valuable components that could be recovered and reused. The typical effluent stream from the nylon industry is a complex mixture that includes n-pentanol, cyclohexanone, and cyclohexene oxide. These chemicals, when separated in their pure forms, have substantial industrial demand. For instance, n-pentanol is used in the production of plasticizers, corrosion inhibitors, and pharmaceuticals, while cyclohexanone and cyclohexene oxide are essential in the production of nylon and as intermediates in other chemical syntheses. Given the significant market for these compounds, their recovery from industrial effluents represents a worthwhile challenge with considerable economic potential. The primary aim of this section is to present the synthesis, design, and evaluation of various separation schemes for the effluent generated by a nylon production facility. The design strategy is based on the application of thermal couplings with the objective of reducing energy consumption. These proposed schemes are evaluated using sustainability metrics and compared to one another to determine the role of thermal couplings in generating sustainable effluent recovery processes (Zhou et al., 2018).

Case study description

In this case study, the separation of an effluent stream generated by the nylon industry is considered. The effluent stream from a nylon production plant contains approximately 500 kg/h of light oil, consisting of n-pentanol (35.4%), cyclohexanone (31.8%), and cyclohexene oxide (32.8%). This mixture, referred to as light oil, does not present azeotropic interactions, making it possible to separate the components using conventional distillation techniques. The target is to achieve purities of at least 95% for each of the components, with a recovery rate exceeding 92.5%. The base case configuration proposed by Zhou et al. (2018) involves a two-column system. In the first column, the light oil is fed into stage 14 of a 30-stage column, with cyclohexanone obtained as the bottom product (see Table 5.2). The remaining mixture, primarily composed of n-pentanol and cyclohexene oxide, is fed into a second column

TABLE 5.2 Design parameter for both conventional schemes.

Scheme	Direct	Direct	Indirect	Indirect
Column	C1	C2	C1	C2
Theoretical stages	60	29	30	70
Diameter (m)	1.2192	0.6096	0.762	0.6096
Feed stage	14	21	14	51
Reflux ratio	68.57	11.9	5.46	23.08
Distillate rate (kg/h)	170.372	165.228	333.955	168.713
Reboiler heat duty (cal/s)	294,220	70,098	65,521	99,702
CO_2 emissions (t/h) × 10^{-1}	1.269		0.5914	
Mass intensity (kg/kg)	1.021		1.023	
TAC (USD/y) × 10^6	1.271		0.905	
Condition number × 10^2	1.16		1.33	
E factor (kg/kg) × 10^{-2}	2.027		2.233	
Inherent safety (probability/y) × 10^{-4}	5.2527		5.2485	

with 70 stages, where cyclohexene oxide is obtained as the top product and n-pentanol as the bottom product. This configuration was simulated in Aspen Plus using the NRTL-HOC thermodynamic model, which is appropriate for the polar, nonelectrolyte nature of the compounds at operating pressures below 10 bar (see Fig. 5.4).

Generation of thermally coupled and intensified configurations

The synthesis of new separation alternatives was performed using the systematic methodology proposed by Errico et al. (2009). The general idea of this methodology is to start from a base conventional configurations (Fig. 5.4) and systematically generate alternatives by introducing thermal couplings, moving column sections, and eliminating sections when possible. This approach ensures that the entire design space is explored to identify configurations that minimize energy consumption while maintaining the desired separation performance (see Fig. 5.5).

Thermally coupled sequences (Fig. 5.5): Starting from the base configuration, thermally coupled sequences were generated by introducing thermal couplings between the condenser of the first column and the reboiler of the second column. This step was intended to allow heat exchange between the two columns, thereby reducing the need for external heating and cooling. Thermally coupled sequences are particularly beneficial because they facilitate direct energy transfer between process units, which leads to significant energy savings. The thermal coupling was implemented in Aspen Plus, and the interconnection flows were optimized using a parametric approach to ensure minimal energy requirements.

Thermodynamically equivalent schemes: Following the introduction of thermal couplings, the next step was to generate thermodynamically equivalent schemes by moving sections within the columns. This involved moving

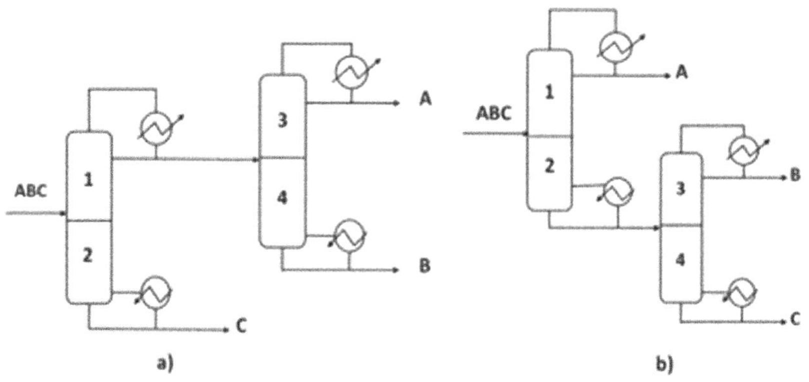

FIGURE 5.4 Conventional configurations for light oil treatment, A for n-pentanol, B for cyclohexanone, C for cyclohexene oxide, (a) indirect sequence, (b) direct sequence.

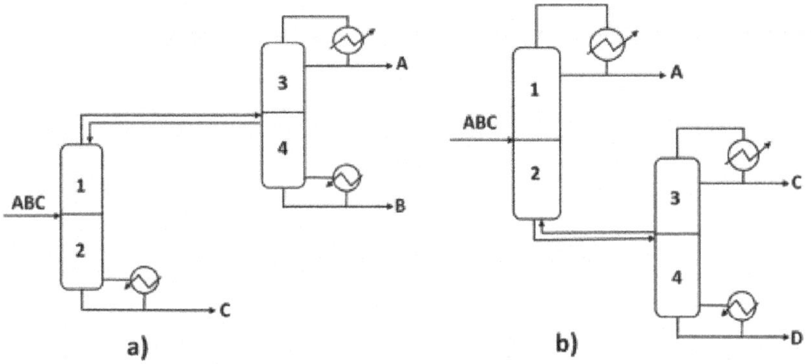

FIGURE 5.5 Thermally coupled configurations for light oil treatment, A for n-pentanol, B for cyclohexanone, C for cyclohexene oxide, (a) indirect sequence, (b) direct sequence.

the rectification or stripping sections of the thermally coupled columns to achieve different configurations with the same thermodynamic efficiency. Thermodynamically equivalent schemes have the same energy requirements as their corresponding thermally coupled designs but may exhibit different operational characteristics, such as improved flexibility or control properties. The thermodynamically equivalent schemes were rigorously simulated in Aspen Plus, and their energy consumption was optimized by adjusting internal flows. The systematic approach ensured that energy savings were achieved while the operational feasibility of the design was maintained (see Fig. 5.6).

Intensified configurations: The intensified configuration was generated by eliminating one of the sections from the thermodynamically equivalent scheme. Specifically, the stripping section of the second column was removed, resulting in a single intensified column capable of performing the separation task of both columns. The removal of a column section results in fewer distillation steps, leading to a decrease in equipment and operational complexity. However, the intensified configuration required higher reboiler duties to achieve the desired separation due to the reduction in equilibrium stages. This configuration was optimized through a parametric approach, focusing on the internal energy distribution to mitigate the increased energy consumption typically associated with intensified designs (see Fig. 5.7).

Dividing Wall Column (DWC): As a final alternative, a dividing wall column was considered. The DWC is an intensified column configuration that integrates the functionality of two or more conventional columns into a single unit with a dividing wall. This setup allows for simultaneous separation of multiple products in a single column shell, which can lead to significant energy savings and reduced equipment costs. The DWC design was derived from the thermodynamically equivalent scheme and optimized in Aspen Plus to determine the optimal liquid and vapor split ratios that minimize energy consumption (see Fig. 5.8).

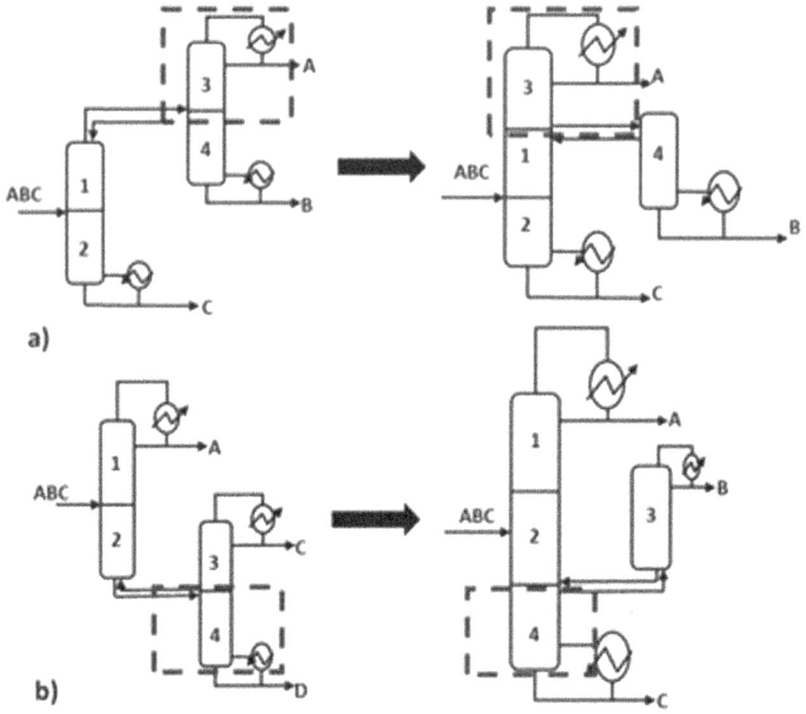

FIGURE 5.6 Thermodynamic equivalent configurations for light oil treatment, A for n-pentanol, B for cyclohexanone, C for cyclohexene oxide, (a) indirect sequence, (b) direct sequence.

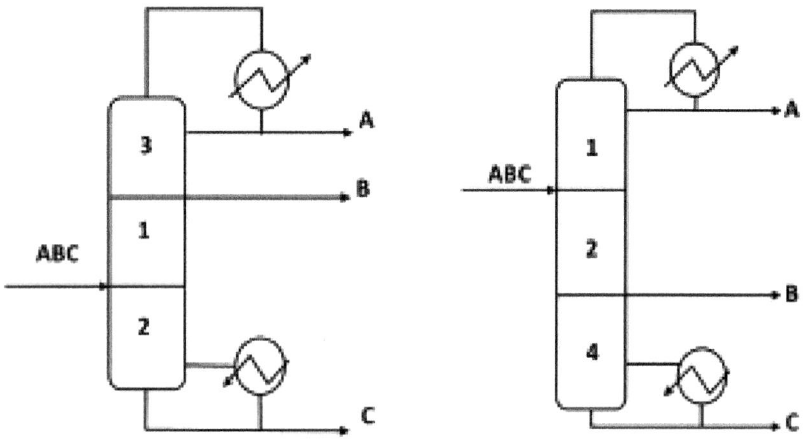

FIGURE 5.7 Intensified configurations for light oil treatment, A for n-pentanol, B for cyclohexanone, C for cyclohexene oxide.

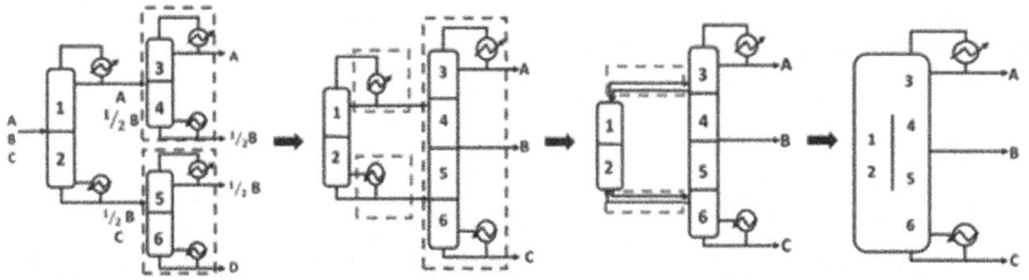

FIGURE 5.8 DWC for light oil treatment, A for n-pentanol, B for cyclohexanone, C for cyclohexene oxide.

Performance indicators for sustainability evaluation

The evaluation of each separation configuration was based on multiple sustainability indicators, each of which provides insights into different aspects of process performance:

Total Annual Cost (TAC): TAC represents the economic viability of each process configuration, considering both capital and operational costs. TAC was calculated as expressed in Chapter 4.

CO_2 emissions: Greenhouse gas emissions were calculated based on the energy consumption of each configuration, providing an environmental indicator of the sustainability of the process. CO_2 emissions were calculated as follow:

$$[CO_2]_{\text{Emissions}} = \left(\frac{Q_{\text{fuel}}}{\text{NHV}}\right)\left(\frac{\%C}{100}\right)\alpha \tag{5.1}$$

Where Q_{fuel} (kW) is the amount of fuel burned, $\alpha = 3.67$ is the ratio of molar masses of CO_2 and C, NHV (kJ/kg) represents the net heating value of a fuel with carbon content of C% (Gadalla et al., 2005).

Mass intensity (MI): MI measures the total mass of materials required to produce a unit mass of product, indicating the resource efficiency of the process. Mass intensity is calculated as follow:

$$\text{MI} = \frac{\text{Mass of all materials used, excluding water (kg)}}{\text{Product mass (kg product)}} \tag{5.2}$$

E-factor: The E-factor represents the total amount of waste generated per unit of product, highlighting the environmental burden associated with each process.

$$\text{E} - \text{Factor} = \frac{\text{Total process waste (kg)}}{\text{kg of product}} \tag{5.3}$$

Condition number: The condition number is a measure of the controllability of the process, with lower values indicating better control properties and easier operational management. Controllability is a crucial consideration for hybrid flowsheets incorporating DWCs. The condition number (CN), derived from the numerical concept of singular value decomposition (SVD), evaluates the sensitivity of the process to disturbances. The SVD of the system matrix K is expressed in Eq. (5.4):

$$K = U\Sigma V^{T} \tag{5.4}$$

Here, K is an n × m matrix representing the process, U and V are orthonormal matrices, and Σ is a diagonal matrix containing singular values ranked as $\sigma_{max} \geq \sigma 1 \geq \ldots \geq \sigma_{min}$. The CN is defined as the ratio of the largest to the smallest singular value (Eq. 5.5):

$$CN = \sigma_{ma}/\sigma_{min} \tag{5.5}$$

A lower CN indicates better controllability and dynamic performance. For each process design, a relative gain matrix is generated under open-loop control conditions by applying a 0.5% positive disturbance to manipulate variables such as reflux ratio, reboiler heat duty, and side stream flow rates. These disturbances are small enough to approximate a first-order system response. To eliminate the scaling dependency of the SVD method, control variables like product purity, naturally bounded between 0 and 1, were paired with appropriately scaled flow rates and manipulated variables. The scaling was performed by assuming that the maximum valve opening is twice the nominal steady-state value, corresponding to a 50% valve position at steady state.

This comprehensive design procedure, integrating economic, environmental, and controllability considerations, ensures that the hybrid flowsheets are not only cost-effective and environmentally sustainable but also robust under operational disturbances.

Inherent safety: Inherent safety was assessed using a quantitative risk analysis approach to estimate the probability of hazardous incidents occurring during operation. Safety is quantified using the Individual Risk Index (IR), which represents the risk faced by an individual based on their specific location. This risk accounts for potential harm, such as injury or death, resulting from an accident with a given frequency of occurrence and a likelihood of impact. Importantly, the IR is independent of the number of individuals exposed. Mathematically, the IR is expressed as the product of the accident frequency (f_i) and the probability of impact at a specific location ($P_{x,y}$), as shown in Eq. (5.6).

$$IR = \sum f_i \quad P_{x,y} \tag{5.6}$$

The frequency and likelihood of impact can be determined through a Quantitative Risk Analysis (QRA), a valuable tool for identifying potential incidents and accidents and assessing their consequences and damages. The initial step in QRA involves identifying possible incidents, defined as the release of matter or energy from a process (Medina-Herrera et al., 2014). For distillation columns, two primary types of incidents are recognized: continuous release, which involves the leakage of material from equipment due to pipeline rupture, and instantaneous release, which refers to the complete loss of material from equipment caused by a catastrophic vessel rupture. These incidents are typically identified using a Hazard and Operability (HAZOP) study.

After identifying likely accidents, the next step is determining the causative variables responsible for injuries or fatalities associated with each accident. For fire-related incidents such as BLEVE, jet fires, and flash fires, the causative variable is thermal radiation (Er) received by a person. For UVCE, the primary cause of fatalities is overpressure (po), while for toxic releases, the causative variable is the concentration (C) of the toxic substance. The methods for calculating these causative variables have been previously detailed by Kumar (1996).

Results and discussion

The results of the sustainability evaluation for the different configurations are summarized in Table 5.3. The thermally coupled configuration demonstrated significant improvements over the base case, reducing energy consumption by approximately 15%. This reduction in energy consumption directly translated into lower operational costs, as well as a decrease in greenhouse gas emissions, with CO_2 emissions reduced by 20% compared to the base configuration. These energy and emission reductions were largely due to the efficient use of thermal coupling, which allowed for the transfer of excess heat from one unit to another without the need for external heating or cooling. The thermodynamically equivalent configuration achieved similar energy savings, with a 16% reduction in energy use. However, it also exhibited better operational flexibility compared to the thermally coupled design, as indicated by the condition number. The condition number for the thermodynamically equivalent scheme was approximately 18% lower than that of the thermally coupled configuration, which indicates improved ease of control. This improvement is particularly important for industrial operations where stability and responsiveness to changes in process conditions are critical for maintaining product quality and safety (Bravo-García et al., 2021).

The intensified configuration, despite eliminating one of the distillation sections, required an increased reboiler duty to achieve the desired separations. Consequently, the energy consumption for this configuration was approximately 5% higher compared to the thermally coupled and thermodynamically equivalent designs. This led to increased operational costs and a slight rise in CO_2 emissions. However, the intensified configuration did provide benefits in terms of equipment costs, reducing TAC by 10% due to the reduction in the number of columns. Moreover, the inherent safety analysis indicated that the

TABLE 5.3 Sustainable indexes for all separation alternatives.

Indicator	Direct	Indirect	TC	TE	Intensified	DWC
CO_2 emissions (t/h) × 10^{-1}	1.269	0.5914	0.55	0.554	3.027	1.073
Mass intensity (kg/kg)	1.021	1.023	1.023	1.023	1.023	1.022
TAC (USD/y) × 10^6	1.271	0.905	0.863	0.866	2.522	1.228
Condition number × 10^2	1.16	1.33	2.44	7.7	2.75	4.6
E factor (kg/kg) × 10^{-2}	2.027	2.233	2.242	2.234	2.252	2.111
Inherent safety (probability/y) × 10^{-4}	5.2527	5.2485	5.4762	5.5097	2.6425	2.6422

simplified process design led to a 15% decrease in the risk of hazardous incidents, which is a crucial consideration for large-scale chemical operations.

The DWC configuration was the most energy-efficient among all alternatives, achieving a 25% reduction in energy consumption compared to the base configuration. The DWC's unique design, which allows for the integration of multiple separation tasks within a single column, enabled significant energy savings and a 22% reduction in CO_2 emissions. However, this design also presented challenges in terms of controllability, as reflected by a higher condition number (25% higher than the thermally coupled scheme). The increased complexity of the DWC, due to the dividing wall and multiple separation regions, made it more challenging to maintain stable operation, particularly during disturbances or changes in feed composition.

In terms of the E-Factor and mass intensity, the thermally coupled, thermodynamically equivalent, and DWC configurations all exhibited significant improvements over the base case. The thermally coupled configuration reduced the E-Factor by 12%, while the DWC achieved a 20% reduction, highlighting the role of thermal coupling in minimizing waste generation. Mass intensity followed a similar trend, with the thermodynamically equivalent configuration achieving a 15% reduction compared to the base configuration. These improvements underscore the importance of thermal couplings not only in enhancing energy efficiency but also in reducing resource consumption and waste production, thus contributing to the overall sustainability of the process.

The analysis of the results demonstrates that thermal couplings play a crucial role in improving both the energy efficiency and environmental sustainability of separation processes in the nylon industry. The thermally coupled and thermodynamically equivalent configurations achieved substantial energy savings and emission reductions while maintaining good controllability and inherent safety characteristics. The DWC, although presenting some challenges in operational control, offered the highest energy efficiency and the greatest potential for reducing greenhouse gas emissions. These findings highlight the potential of thermal coupling and process intensification strategies to contribute to the sustainable development of the chemical industry, particularly in the context of effluent recovery and circular economic principles.

Use of thermal couplings in the production processes of compounds with renewable raw materials

The growing focus on sustainability, propelled by international policies such as the United Nations' 2030 Agenda for Sustainable Development, has prompted a fundamental shift in the chemical industry (Biermann et al., 2017). The 2030 Agenda emphasizes the necessity of responsible production and consumption (SDG 12), climate action (SDG 13), and the transition toward a circular economy. A key element of the circular economy is the minimization of waste and the efficient use of resources, where processes are designed to recover, recycle, and reintegrate materials back into the production cycle (Ogunmakinde et al., 2022). This shift is increasingly driving industries to adopt renewable feedstocks, such as biomass, for the production of high-value-added chemicals, which offers an alternative to the traditional dependence on fossil-based raw materials. The use of renewable raw materials not only reduces the environmental footprint of chemical processes but also aligns with the principles of a sustainable and circular economy. Producing high-value compounds from biomass presents an opportunity to embed chemical production into a sustainable framework, where renewable resources are used to generate chemicals that are fundamental to the production of polymers, fuels, and other essential industrial materials. Biomass-derived compounds like 2,3-butanediol and levulinic acid are prime examples of valuable platform chemicals that can be produced sustainably and serve as precursors for a wide range of high-value products (Garg & Jain, 1995). The production of these chemicals from renewable feedstocks offers a promising route to achieving carbon neutrality, as they can help offset greenhouse gas emissions and contribute to reducing the overall carbon footprint of the chemical sector (Celińska & Grajek, 2009).

However, the use of renewable feedstocks in chemical production introduces specific challenges related to process efficiency and energy consumption. Biomass-derived feedstocks often contain a variety of impurities, require additional pretreatment steps, and tend to have complex chemical compositions that make their separation more challenging compared to conventional petrochemical processes. Addressing these challenges requires innovative approaches that can enhance process efficiency while maintaining sustainability goals. In this context, thermal couplings emerge as a powerful tool to improve the energy efficiency of separation processes. Thermal couplings enable efficient heat integration across different units, reducing the overall energy required for separation and thus decreasing operational costs and environmental impact. By enhancing energy efficiency and integrating heat utilization, thermal couplings directly support the sustainability objectives outlined in the 2030 Agenda, particularly by contributing to reduced energy consumption and minimizing greenhouse gas emissions. In biomass valorization processes, thermal couplings can play an essential role by making complex separation steps more efficient. For example, the purification of 2,3-butanediol and levulinic acid—compounds that are typically produced through fermentation or chemical conversion of biomass—requires extensive separation processes to achieve the desired purity levels. Incorporating thermal couplings into these processes allows for significant energy savings, which not only enhances economic feasibility but also aligns with environmental and sustainability goals. As such, the use of thermal couplings in these contexts helps bridge the gap between the theoretical sustainability potential of biomass and its practical application in industrial-scale production (Song et al., 2017).

In the following sections, we will explore specific case studies of the use of thermal couplings in the purification of 2,3-butanediol and levulinic acid. These case studies will illustrate how the integration of thermal couplings can lead to significant improvements in process performance, energy savings, and overall sustainability when dealing with renewable raw materials. The focus will be on understanding how these technologies contribute to sustainable chemical production and align with current global policies on sustainability and the circular economy.

Case study: 2,3-butanediol purification

2,3-Butanediol (2,3-BD) is a valuable platform chemical with a wide range of industrial applications. It is a diol with the chemical formula $C_4H_{10}O_2$, and it exists as three stereoisomers: meso-2,3-butanediol, D-(+)-2,3-butanediol, and L-(−)-2,3-butanediol. This compound is primarily used as an intermediate for the production of chemicals such as methyl ethyl ketone (MEK), butadiene, and diacetyl, all of which are important in the manufacture of plastics, rubber, solvents, and various specialty chemicals. In addition, 2,3-BD is used in the food industry as a flavor enhancer and in the pharmaceutical sector for synthesizing drugs and other bioactive compounds. Conventionally, 2,3-BD is produced via petrochemical processes that involve the catalytic hydration of butadiene or the oxidation of propylene. These methods, while effective, rely heavily on nonrenewable fossil-based feedstocks, which contributes to greenhouse gas emissions and environmental degradation. In light of the increasing focus on sustainability and the circular economy, alternative production routes that are more environmentally friendly are being actively pursued (Wu et al., 2008).

One of the most promising sustainable routes for the production of 2,3-BD is through microbial fermentation. In this process, various microbial strains, such as *Klebsiella pneumoniae*, *Bacillus licheniformis*, and *Enterobacter aerogenes*, are used to convert biomass-derived carbohydrates into 2,3-BD under anaerobic conditions. The feedstocks for these processes are typically renewable, such as glucose, sucrose, or agricultural residues, making this approach a more sustainable alternative compared to traditional petrochemical routes (Kooi et al., 1948). However, despite its sustainability benefits, fermentation-based production of 2,3-BD faces several challenges, primarily related to the low concentration of 2,3-BD in the fermentation broth. Typically, the concentration of 2,3-BD in fermentation broth ranges from 20 to 40 g/L, which is relatively low compared to other fermentation products. This low concentration results in high downstream processing costs, which limit the economic viability of the process (Dai et al., 2015). Nevertheless, there are some advantages to the composition of the fermentation broth. Unlike many other bio-based fermentation products, the fermentation mixture containing 2,3-BD is relatively straightforward to separate because it does not form azeotropes or exhibit complex thermodynamic interactions. As a result, conventional separation techniques such as distillation can be considered a viable option for purifying 2,3-BD from the fermentation broth (Song et al., 2019). However, distillation is an energy-intensive process, and there is a strong incentive to improve its efficiency by employing advanced process design strategies.

Case study description

Despite the thermodynamic advantages of the fermentation mixture, the separation of 2,3-BD has not been extensively explored. This study takes as its basis the proposal presented by Penner et al. (2017), which consists of a sequence of direct and indirect distillation columns. This configuration will serve as the foundation for developing more advanced separation alternatives. In this study, the feed considered consists of a fermentation broth containing approximately 500 kg/h of 2,3-BD, with a composition of 35% water, 30% 2,3-butanediol, and 35% other minor fermentation by-products. The goal is to achieve a purity of at least 99% for 2,3-BD, with a recovery rate exceeding 95%. The separation process will be modeled using Aspen Plus software, employing the NRTL (Non-Random Two-Liquid) thermodynamic model due to its suitability for highly nonideal mixtures containing polar compounds. The designs will be evaluated in terms of energy consumption, product purity, and overall process feasibility.

The synthesis and design methodology employed in this study builds upon the base distillation sequences proposed by Penner et al. (2017). Initially, a conventional direct and indirect distillation sequence was modeled, which served as the starting point for exploring advanced separation schemes (see Fig. 5.9). The subsequent steps involved applying systematic process intensification and thermal coupling methodologies, as previously explained by Errico et al. (2009). These advanced configurations include thermally coupled, thermodynamically equivalent, and intensified schemes, each designed to enhance energy efficiency and improve overall process sustainability.

Considering the base sequences, the systematic methodology previously outlined by Errico et al. (2009) was applied to propose thermally coupled, thermodynamically equivalent, and intensified schemes (see Fig. 5.10). Thermally coupled sequences involve the introduction of heat integration between different distillation columns, enabling more efficient energy use by reducing the need for external heating and cooling. Thermodynamically equivalent schemes are generated by rearranging sections of the thermally coupled columns to achieve the same thermodynamic performance but potentially improve operational flexibility or control properties. Intensified schemes involve the removal of certain column sections to simplify the

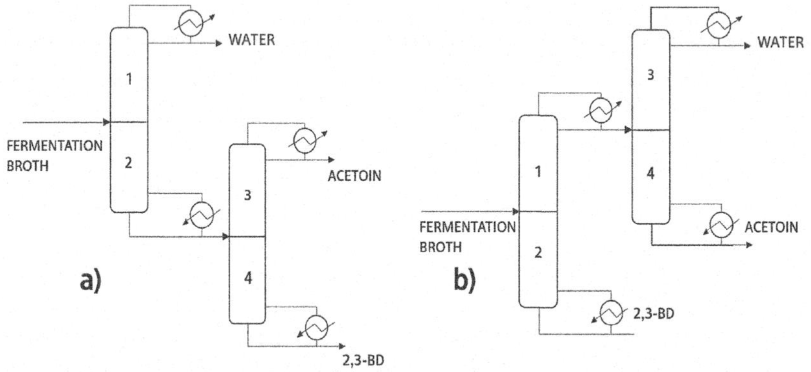

FIGURE 5.9 Base sequences for 2,3-BD purification; (a) direct sequence, (b) indirect sequence.

process, reduce equipment count, and decrease capital investment, albeit sometimes at the expense of increased reboiler duties.

To evaluate the sustainability of each design, a multi-objective optimization framework was utilized. The optimization was performed using (DETL), which is a robust optimization algorithm that combines the exploration capabilities of differential evolution with a memory-based approach to prevent revisiting previously explored solutions. This hybrid approach enables a thorough search of the solution space to identify optimal trade-offs between different objectives (Sharma & Rangaiah, 2013).

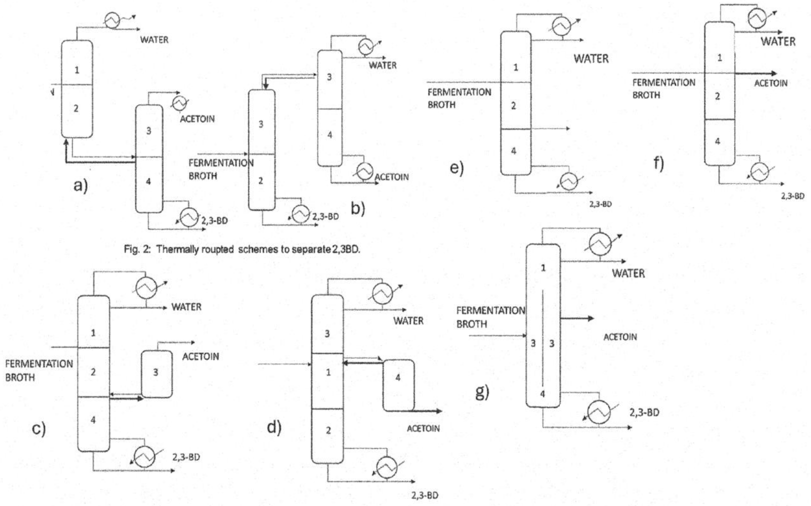

FIGURE 5.10 Separations alternatives for 2,3-BD purification; being (a and b) thermally coupled, (c and d) thermodynamic equivalent, (e and f) intensified, (g) DWC.

In this study, the DETL optimization focused on minimizing three key sustainability indicators: Total Annual Cost (TAC), Eco-indicator 99 (EI99), and Individual Risk Index (IR). The TAC represents the economic viability of each configuration, accounting for both capital and operational expenses, including energy costs. The Eco-indicator 99 (EI99) is used as an environmental metric, evaluating the overall environmental burden associated with each design, including resource consumption and greenhouse gas emissions throughout the process lifecycle. The Individual Risk Index (IR) provides a measure of the inherent safety of each process, quantifying the probability of hazardous incidents that could occur during operation. By employing DETL, a comprehensive exploration of the process design space was carried out to determine the most effective balance between economic, environmental, and safety performance. This allowed the generation of a Pareto front that represents the optimal trade-offs between the three objectives, highlighting designs that achieve significant energy and cost reductions while also enhancing safety. The inclusion of a tabu list in the optimization algorithm ensures that previously explored solutions are not revisited, thus improving the efficiency of the search process and ensuring a diverse range of potential solutions.

Results and discussion

The results of this study revealed significant improvements when comparing the base configuration to the thermally coupled and thermodynamically equivalent alternatives. Specifically, the thermally coupled sequence achieved a 20% reduction in energy consumption, which directly translated into a decrease in TAC by 15% compared to the base configuration. The thermally coupled configuration also demonstrated a 25% reduction in EI99, highlighting the role of thermal couplings in enhancing environmental sustainability through improved energy efficiency. In addition, the Individual Risk Index for the thermally coupled configuration was approximately 18% lower than that of the base case, indicating a safer process design due to reduced equipment usage and better heat management. Table 5.4 shows the objective function for all the alternatives studied in this section.

The thermodynamically equivalent configuration exhibited similar energy savings but showed enhanced controllability, as indicated by a 12% reduction in the condition number compared to the thermally coupled scheme. This improved operational flexibility can be critical in ensuring stable operation under variable process conditions. The intensified configuration, while providing some advantages in terms of equipment count reduction, required higher reboiler duties, resulting in a slight increase in TAC and a marginal rise in greenhouse gas emissions. However, the inherent safety risk associated with the intensified configuration was reduced by 10%, owing to fewer potential failure points in the simplified design. Table 5.5 shows the main design parameters for some schemes in Fig. 5.10.

TABLE 5.4 Objective function values for all separation alternatives.

Objective function	Direct (Fig. 5.9a)	Indirect (Fig. 5.9b)	Thermally coupled (Fig. 5.10a)	Thermodynamic Equivalent (Fig. 5.10c)	Intensified (Fig. 5.10e)	DWC (Fig. 5.10g)
TAC($/y)	35,032,419	51,155,609	31,360,313	313,055,124	30,536,031	231,268,638
EI99 (points/y)	14,328,558	22,627,903	12,559,191	12,557,857	12,407,199	97,659,879
IR (probability/y)	0.0006686	0.0006663	0.0006795	0.0006684	0.00033411	0.0003661

TABLE 5.5 Design parameters for thermally coupled, thermodynamic equivalent, intensified, and DWC schemes.

Configuration		Number of theoretical stages	Reflux ratio	Feed stage	Column diameter (m)	Column height (m)	Distillate flowrate (kmol/h)	Interconnection vapor flow (kmol/h)	Interconnection liquid flow (kmol/h)	Wall location	Side stage	Side flow (kmol/h)	Reboiler duty (kW)
Thermally coupled	C1	49	0.3	42	0.675	28.7	3417.5	–	3986.76	–	–	–	–
	C2	50	68.5	12	0.502	29.3	7.62	3859.5	–	–	–	–	65,861
Thermodynamic equivalent	C1	87	0.3	42	0.675	51.8	3416.28	–	32	–	–	–	64,439
	C2	12	25.023	12	0.502	6.1	7.14	24.85	–	–	–	–	–
Intensified	C1	87	0.333	42	0.675	51.8	3416.07	–	–	–	49	8.155	61,457
DWC	Section 4	15	–	12	0.942	7.9	–	–	–	32–36	–	–	–
	Sections 1–3	71	10.09	–	1.493	42.1	3415.85	–	–	–	50	8.084	438,198

These findings underscore the critical role of thermal couplings in achieving sustainable process designs. By facilitating effective heat integration, thermal couplings not only reduce energy requirements but also enhance the economic feasibility and safety of the 2,3-BD purification process. The DETL optimization approach played a crucial role in identifying configurations that achieve an optimal balance between economic, environmental, and safety objectives. This comprehensive approach to separation design is essential to address current industrial demands for sustainable and economically viable chemical production.

Case study: Levulinc acid purification

Levulinic acid (LA) is a platform chemical of significant industrial importance, often recognized for its versatility and potential as a renewable building block for various high-value products. As an organic compound derived from the degradation of cellulose, LA plays a crucial role in the production of a variety of downstream chemicals, including fuel additives, solvents, plasticizers, and pharmaceutical intermediates. Its relevance in today's society is amplified by its alignment with the principles of green chemistry and sustainability, making it an integral part of the drive toward developing renewable and environmentally friendly chemical production routes. The production of levulinic acid from biomass, such as lignocellulosic feedstocks, is a promising approach that aligns with current sustainability goals, as it enables the utilization of waste biomass to generate valuable products, thus promoting a circular economy and minimizing environmental impact (Boddula et al., 2023).

However, one of the major challenges associated with the production of levulinic acid from biomass is the complexity of the purification process due to the intricate thermodynamic interactions in the mixture. The presence of azeotropes in the effluent stream complicates the separation process, making conventional separation methods less efficient. The azeotropic nature of the mixture, which typically includes components such as water, levulinic acid, and by-products like formic acid and furfural, introduces significant challenges in terms of energy requirements and process feasibility (Alcocer-García et al., 2019). In this context, the objective of this section is to present several separation alternatives generated through the use of thermal couplings, with the aim of achieving substantial energy savings. The methodology applied here follows a similar approach to that used in the previous sections; however, given the larger number of components involved in the feed mixture, the methodology was extended to include a greater number of columns, thus expanding the scope of the process synthesis and intensification efforts.

Case study description

In this case study, the purification of levulinic acid from a biomass-derived effluent is considered. The feed for the purification process consists of a mixture of approximately 1000 kg/h, containing 45% water, 30% levulinic acid, 15% formic acid, and 10% furfural. The complexity of this feed lies in the presence of azeotropic interactions among the components, particularly between water and formic acid, as well as between water and furfural, which significantly complicates the separation process (see Fig. 5.11). The goal of this study is to achieve a purity of at least 99% for levulinic acid, while maximizing recovery and minimizing energy consumption (Alcocer-García et al., 2019).

The process was modeled using Aspen Plus, with the NRTL-HOC (Non-Random Two-Liquid-Hayden O'Connell) thermodynamic model employed to accurately capture the nonideal behavior of the azeotropic mixtures involved. The base configuration for the purification process consists of an initial liquid–liquid extraction column, followed by three distillation columns for further purification of the levulinic acid and the separation of the by-products. This base case serves as the starting point for the synthesis of alternative configurations (see Fig. 5.12).

Following the synthesis of the base case, several advanced separation alternatives were generated by introducing thermal couplings between the columns. The thermal coupling approach allows for the efficient exchange of heat between different parts of the process, reducing the overall energy requirements and thus improving the sustainability profile of the process. The synthesis methodology was extended to accommodate a greater number of distillation columns, reflecting the increased complexity of the mixture compared to previous case studies. The alternatives included configurations

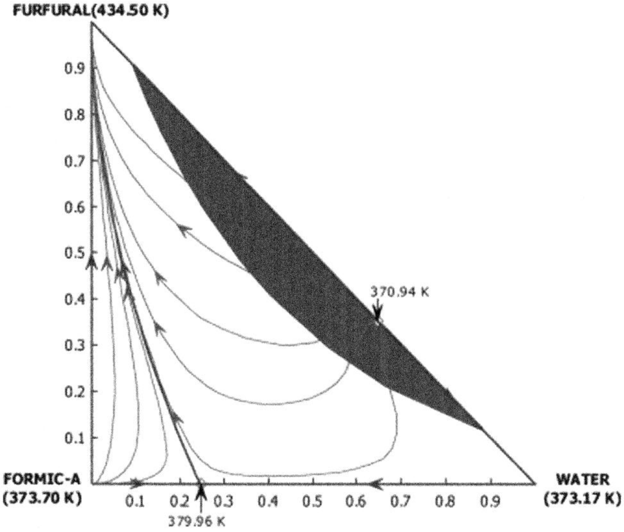

FIGURE 5.11 Azeotropes of the mixture furfural-water and formic acid.

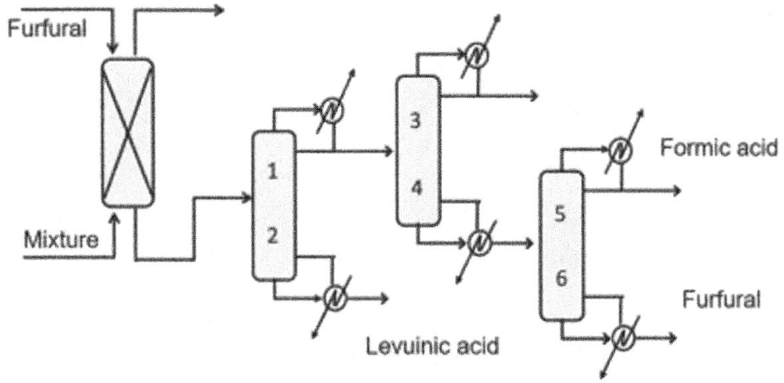

FIGURE 5.12 Conventional sequence (CS).

with multiple thermal couplings between the distillation columns, as well as thermodynamically equivalent configurations generated by repositioning column sections to achieve similar thermodynamic efficiency. See through Fig. 5.13 the entire methodology to generate separation alternatives.

Once the synthesis work was completed, a multi-objective optimization was conducted to evaluate all the generated alternatives. The optimization considered both economic and environmental objectives, specifically focusing on minimizing the Total Annual Cost (TAC) and the Eco-indicator 99 (EI99), which serves as an environmental metric to quantify the impact of each configuration throughout its life cycle. DETL was employed for the optimization, allowing for an efficient exploration of the design space and identification of optimal trade-offs between economic and environmental performance (Sharma & Rangaiah, 2013).

Results and discussion

The results from the sustainability evaluation and optimization of the different configurations demonstrated significant improvements in the thermally coupled alternatives compared to the base case (see Table 5.6). Specifically, the thermally coupled sequences achieved energy savings of up to 25%, which directly translated into a reduction in the TAC by approximately 20% relative to the base configuration. This reduction in energy consumption and cost was primarily attributed to the enhanced heat integration facilitated by the thermal couplings, which reduced the reliance on external heating and cooling utilities and thus improved the overall energy efficiency of the process.

In addition to the economic improvements, the thermally coupled configurations also showed substantial environmental benefits, with reductions of up to 30% in the Eco-indicator 99 compared to the base case. The reduction in EI99 indicates that the thermally coupled schemes not only consume less energy but also generate fewer emissions and have a lower overall

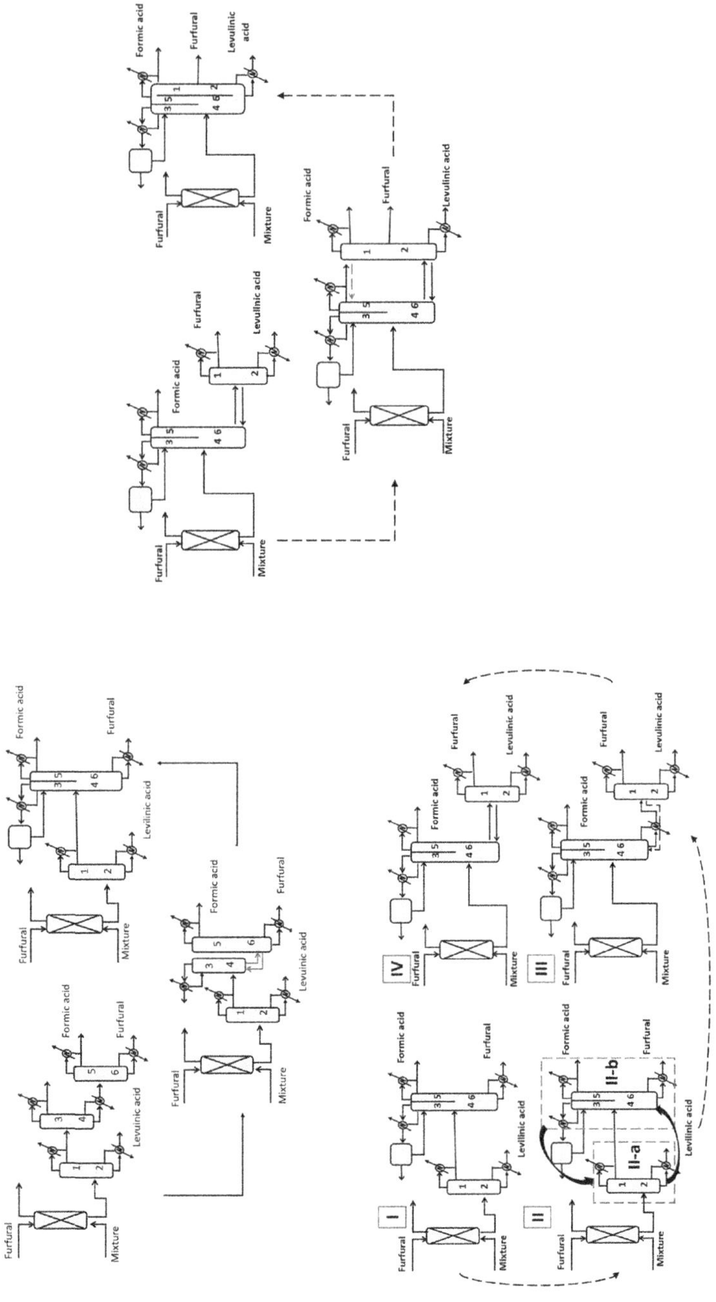

FIGURE 5.13 Synthesis of the dividing wall column configuration with a decanter and thermal coupling.

TABLE 5.6 Design parameters and performance indexes for CS schemes.

	CS1	CS2	CS4
Number of stages	22	24	28
Extractant (kg/h)	106.722	106.154	106.018
Diameter (m)	1.434	0.538	0.987
Number of stages	28	24	17
Feed stage	13	16	11
Reflux (kg/kg)	0.076	0.025	0.026
Reboiler duty (MW)	23.026	10.278	9.403
Diameter (m)	3.642	0.653	1.276
Number of stages	46	39	35
Feed stage	34	32	9
Reflux (kg/kg)	0.206	64.901	0.125
Reboiler duty (MW)	4.465	3.311	13.935
Diameter (m)	2.413	0.431	0.983
Number of stages	60	39	56
Feed stage	28	13	16
Reflux (kg/kg)	54.199	0.11	43.9
Reboiler duty (MW)	3.486	13.571	3.237
Diameter (m)	2	0.512	1.662
Levulinic acid (%w/w)	0.98	0.98	0.981
Formic acid (%w/w)	0.851	0.85	0.857
Furfural (%w/w)	0.999	0.999	0.999
TAC ($/y)	1.85E+07	1.56E+07	1.49E+07
EI99 (points/y)	5.36E+09	4.70E+09	4.60E+09
TAC saving	–	16%	19%
EI99 saving	–	12%	14%

environmental impact throughout their life cycle (Goedkoop & Spriensma, 2001). The role of thermal couplings was particularly significant in reducing energy-intensive reboiler duties, thereby contributing to lower greenhouse gas emissions and a smaller carbon footprint. The thermodynamically equivalent configurations, while offering similar energy savings to the thermally coupled schemes, demonstrated improved controllability, as indicated by a lower condition number. This suggests that these configurations may offer better operational stability, which is critical in ensuring consistent product quality and maintaining process safety, especially in the presence of disturbances. However, the intensified configurations, which involved the removal of some column sections, presented a trade-off: while they resulted in a simplified process with fewer equipment requirements, they also led to slightly increased reboiler duties and a marginal increase in TAC and environmental impact.

Overall, the results highlight the critical role of thermal couplings in achieving sustainable process designs for the purification of levulinic acid. By facilitating effective heat integration, thermal couplings not only reduce energy consumption and costs but also significantly improve the environmental performance of the process. These findings demonstrate the potential of using advanced process synthesis and optimization techniques to meet the current demands for sustainable and economically viable chemical production processes.

Evaluation of dynamic properties of thermally coupled schemes

Controllability plays a crucial role in the operation of chemical processes, ensuring that a process is maintained at desired setpoints despite disturbances that may arise from changes in feed composition, equipment performance, or other external factors. It is necessary to maintain a controlled process within specific setpoints to guarantee that the process remains efficient, safe, and economically viable. A process that lacks control is more susceptible to deviations that can lead to off-specification products, increased waste, safety risks, and higher operational costs. Therefore, the ability to effectively control a chemical process is fundamental to achieving its operational objectives and ensuring long-term sustainability (Stephanopoulos, 1984, p. 502–504).

From a sustainability perspective, controllability is essential for minimizing the environmental impact of a chemical process. When a process is well-controlled, it operates closer to optimal conditions, which reduces waste generation and energy consumption. Poorly controlled processes can result in products that do not meet quality specifications, requiring reprocessing or disposal, which in turn leads to increased raw material usage, energy consumption, and environmental emissions. By maintaining product quality within a consistent range, controllability directly contributes to reducing the environmental footprint of the process, thus promoting sustainable chemical production (Luyben, 2013).

Controllability is also closely related to process safety. Chemical processes often involve hazardous materials and operate under extreme conditions of temperature and pressure, which can pose significant safety risks if not properly managed. A well-controlled process is inherently safer because it is less likely to deviate from safe operating conditions, thereby minimizing the risk of incidents such as runaway reactions, equipment failures, or unintentional releases of hazardous substances. Effective control strategies ensure that key variables, such as temperature, pressure, and flow rates, remain within safe limits, preventing conditions that could compromise process safety (Crowl & Louvar, 2011).

Additionally, maintaining consistent control allows the process to operate efficiently, ensuring that products are always produced within the specified quality range. This reduces the need for additional reprocessing or disposal of substandard products, thus avoiding unnecessary waste generation and maximizing resource efficiency. Ultimately, the strong link between controllability, safety, efficiency, and waste minimization underscores its importance in sustainable chemical production.

In Chapter 4, the synthesis, design, and simulation of thermally coupled schemes for the separation of biobutanol were discussed. These designs were evaluated in a steady-state context, meaning that the analysis considered all variables to be constant over time, without accounting for dynamic variations. The steady-state evaluation demonstrated the positive impact of thermal couplings on the performance of butanol separation alternatives, highlighting significant energy savings and improved economic performance. However, the steady-state analysis does not capture the dynamic behavior of the processes, which is critical to understanding how they perform under real-world conditions where variables are constantly changing.

The dynamic performance of a process is crucial for sustainability, as it directly influences how well the process can handle disturbances and maintain stable operation. A process that has poor dynamic characteristics may require frequent adjustments or additional energy inputs to maintain the desired product quality, which can increase operational costs and environmental emissions. Therefore, evaluating the dynamic behavior of thermally coupled schemes is essential to assess their viability from a sustainability perspective. Controllability is a key factor that can significantly impact both the economic and environmental performance of the process under dynamic conditions (Grossmann & Daichendt, 1996).

The objective of this section is to evaluate the dynamic properties of the thermally coupled schemes for biobutanol separation using two different control strategies: open-loop and closed-loop control. In an open-loop strategy, the process operates without feedback control, relying on fixed setpoints and operating conditions. While this strategy is not used in the industry as a control mechanism due to the absence of controllers, it provides valuable theoretical insights into the natural dynamics of the process in the presence of disturbances. Open-loop analysis allows for a qualitative understanding of the process dynamics, serving as a benchmark to evaluate how the system responds without intervention. On the other hand, closed-loop control actively monitors process variables and adjusts the operating parameters to maintain the desired setpoints, even in the presence of disturbances. Closed-loop control is therefore essential for ensuring stability, safety, and sustainability in chemical processes (Smith & Corripio, 2005, p. 13–68).

The dynamic analysis will involve evaluating the response of each thermally coupled scheme to various disturbances, such as changes in feed flow rate, feed composition, and operating temperatures. Key performance indicators such as settling time, overshoot, and deviation from setpoint will be used to assess the effectiveness of each control strategy in maintaining stable operation. The comparison between open-loop and closed-loop control will provide insights into the dynamic properties of the thermally coupled schemes and highlight their potential for sustainable operation under real-world conditions.

Open-loop control analysis

Open-loop control in chemical processes refers to the system's natural response to disturbances without the intervention of a feedback control mechanism. In an open-loop configuration, the input is applied to the system, and the resulting output is observed without making any adjustments based on the output itself. This approach, while not used in industry for maintaining stability under real-time conditions, is extremely valuable from a theoretical perspective as it allows for the analysis of the natural dynamics of a system. Understanding the inherent response characteristics of a process under open-loop conditions provides insights into stability, damping, and time constants, which are essential for the proper design of effective control systems (Klema & Laub, 1980).

One of the useful tools in analyzing open-loop control dynamics is Singular Value Decomposition (SVD). The SVD provides a method to decompose a matrix into three separate matrices that reveal important properties about the system. For a process with transfer function matrix G(s), the SVD can be used to represent it in the form:

$$G(s) = U(s)\Sigma(s)V(s)^T \tag{5.7}$$

Where U(s) and V(s) are orthogonal matrices representing the input and output spaces, respectively; and is a diagonal matrix containing the singular values.

The singular values, denoted as σ_i, provide valuable information about the system's gain characteristics in different directions of the input and output spaces. By analyzing these values, we can understand the system's sensitivity to inputs and identify directions where the system may be highly responsive or insensitive. The largest singular value represents the direction in which the system has the highest gain, and thus is most sensitive to disturbances. Conversely, the smallest singular value corresponds to the direction of least sensitivity.

The ratio of the largest singular value to the smallest singular value, called the condition number, provides insight into the system's robustness and sensitivity to disturbances. A high condition number indicates that the system is highly sensitive and may have poor controllability, making it susceptible to disturbances. Therefore, SVD analysis is a critical tool for assessing the open-loop behavior of thermally coupled processes, providing information that can be used to improve system controllability (Moore, 1986).

Closed-loop control analysis

Closed-loop control systems are employed to mitigate the impact of disturbances by adjusting process variables in response to changes in the system's output. In a closed-loop configuration, sensors monitor the process variables, and feedback controllers make real-time adjustments to maintain the desired setpoints. This approach is fundamental to maintaining process stability, achieving target performance, and ensuring safety in chemical processes. Closed-loop control is particularly crucial in thermally coupled schemes where multiple variables interact, and maintaining system stability can be challenging without proper feedback mechanisms (Asbjornsen, 1985).

Types of controllers

The most common types of controllers used in chemical processes are Proportional (P) and Proportional-Integral (PI) controllers. These controllers are used to regulate various process variables, such as temperature, pressure, and flow rate, by adjusting the manipulated variables based on the deviation (error) between the measured output and the desired setpoint.

Proportional (P) Controller: The output of a proportional controller is proportional to the error (Eq. 5.8). It is represented mathematically as:

$$u(t) = K_p e(t) \tag{5.8}$$

Where u(t) is the control action, K_p is the proportional gain, and e(t) is the error between the setpoint and the process variable. The P controller provides a fast response but often results in steady-state errors.

Proportional-Integral (PI) Controller: The PI controller not only responds to the current error but also integrates the error over time to eliminate steady-state error (Eq. 5.9). It is represented as:

$$u(t) = K_p e\ (t) + K_i \int_0^t e(\tau)\, d\tau \tag{5.9}$$

Where is the integral gain. The PI controller is widely used in chemical processes due to its ability to eliminate steady-state error while providing good control performance (Luyben, 2013).

Performance metrics for closed-loop control

The performance of a closed-loop control system can be evaluated using several metrics that quantify how well the system is able to maintain the desired setpoint in the presence of disturbances. The most common performance metrics are:

Integral of Absolute Error (IAE): This metric measures the accumulated absolute error over time (Eq. 5.10):

$$\mathrm{IAE} = \int_0^T |e(t)|dt \tag{5.10}$$

A lower IAE value indicates that the system can minimize deviations from the setpoint over time, providing better performance.

Integral of Squared Error (ISE): The ISE metric is defined as (Eq. 5.11):

$$\mathrm{ISE} = \int_0^T e(t)^2 dt \tag{5.11}$$

ISE penalizes larger errors more heavily, making it useful for identifying systems with significant deviations from the setpoint.

Integral of Time-weighted Absolute Error (ITAE): ITAE places greater emphasis on errors that persist for a long time, thereby encouraging faster settling times (Eq. 5.12):

$$\mathrm{ITAE} = \int_0^T t|e(t)|dt \tag{5.12}$$

These performance metrics provide quantitative measures for evaluating the effectiveness of different control strategies, helping to ensure that the process remains stable, efficient, and within desired operating limits (Luyben, 1992).

Tuning of controllers

Proper tuning of controllers is essential for achieving optimal closed-loop performance. Controller tuning involves adjusting the controller parameters to achieve a balance between response speed, stability, and robustness. A well-tuned controller provides a rapid response to disturbances while minimizing overshoot and oscillations.

Several tuning methods are available, including Ziegler-Nichols tuning and Cohen-Coon tuning, both of which provide initial parameter estimates that can be further fine-tuned through simulation or experimentation (Ziegler-nichols, 2012, p. 1–16). The tuning process is critical for ensuring that the closed-loop control system can effectively respond to disturbances and maintain the desired setpoints.

Case study: Biobutanol purification

In this chapter, both open-loop and closed-loop analyses will be used to evaluate the dynamic properties of the thermally coupled schemes for biobutanol separation. In the open-loop analysis, the natural dynamic response of each scheme will be examined to identify potential issues related to stability, sensitivity, and responsiveness. The singular value decomposition (SVD) approach will be applied to assess the inherent controllability of the system and identify directions in which the system is most sensitive to disturbances.

The closed-loop analysis will involve implementing PI controllers to regulate key process variables. The controllers will be tuned to achieve the desired dynamic performance, and their effectiveness will be assessed using metrics such as IAE. By comparing the closed-loop responses of the different thermally coupled schemes, insights will be gained into the impact of thermal couplings on process controllability, stability, and overall performance.

The evaluation of dynamic properties is critical to understanding how thermally coupled schemes perform in real-world applications where disturbances and variability are inevitable. The results of this dynamic analysis will help identify the most sustainable and robust configurations for biobutanol separation, ensuring that the process can maintain stable and efficient operation under changing conditions.

To further analyze the dynamic performance of the separation schemes for biobutanol, both open-loop and closed-loop control strategies will be applied to study the case presented in Chapter 4, which focused on the purification of biobutanol. Specifically, this analysis will be conducted on four distinct separation alternatives: the best conventional design, the best thermally coupled scheme, the best thermodynamically equivalent scheme, and the best intensified configuration with fewer than $N-1$ columns. All of these separation alternatives involve hybrid processes that combine distillation and liquid-liquid extraction. The feed stream for these processes consists of a mixture of biobutanol and other fermentation by-products. The goal is to evaluate the dynamic behavior of these hybrid configurations to determine their controllability and robustness in response to process disturbances.

The dynamic analysis will build upon the steady-state evaluation presented in Chapter 4, where the Total Annual Cost (TAC) and environmental impact were assessed for each of the separation alternatives. By evaluating the controllability of each scheme, the current analysis aims to provide a more comprehensive understanding of the overall sustainability of the separation designs. The ability to maintain stable and efficient operation under dynamic conditions is an important factor in ensuring the sustainability of the separation schemes, as it directly influences operational efficiency, energy consumption, safety, and waste generation. By combining the insights gained from steady-state and dynamic analyses, this study will provide a rounded assessment of the sustainability of the separation alternatives for biobutanol purification. The objective is to identify designs that not only minimize economic and environmental impacts but also exhibit robust dynamic behavior, thereby contributing to the development of truly sustainable chemical processes.

In this part of the chapter, the reader will be guided step-by-step on how to perform this dynamic analysis using Aspen Plus. Aspen Plus is a powerful process simulation tool that can be used to model both the steady-state and dynamic behavior of chemical processes. The dynamic modeling of the thermally coupled schemes will involve creating a process flowsheet in Aspen Plus, defining all process streams, unit operations, and control loops. The analysis will then involve running both open-loop and closed-loop simulations to assess the system's response to various disturbances.

Study of open-loop control properties in a simulator environment

Initially, it is necessary to have a steady-state Aspen Plus simulation that has been properly converged. In this case, since the chapter is focused on distillation columns with thermal couplings, a distillation column will be considered as a quasi-study. However, the procedure is relatively similar for other modules within the aspen plus simulator. Thus, once the simulation has converged, it is necessary to perform some preliminary steps before the simulation can be "exported" to the dynamic state.

It is necessary that a pressure drop profile has been established for the column we are working with, either by indicating the dome and bottom pressure or by indicating a pressure profile accordingly. Also, the column in question needs to have a value for the column diameter assigned to it; either obtained by using the "tray sizing" tool or calculated by some other methodology. Finally, in the "Setup" menu in the 'Specifications' section, it is necessary to select the "Dynamic" option in the "input mode" menu in the "Global settings" section. Fig. 5.14 shows graphically the above instructions, depending on the version of Aspen Plus, the interface may vary slightly.

Once the change from steady state to dynamic state has been made, it will be necessary to complete some additional inputs into the simulator. In order to simulate the internal and external dynamics of the columns, the decanter, the reflux drum as well as the sump must be sized. To size the vessels, it is necessary to specify the dimensions in the 'Dynamics' menu on the left side of Aspen (5 ft length and 3 ft diameter is a good rule of thumb). Also, it is possible to immediately locate the missing inputs by pressing the "Next" button. When the equipment has been sized, the "Rigorous" calculation of the column hydraulics must be specified in the "Hydraulics" menu, located in the same section where the reflux drum and sump were sized. Once the inputs are properly placed, it is possible to export the simulation to dynamic; for this, it is necessary to select the "File" menu of Aspen Plus, select the "Export" option,

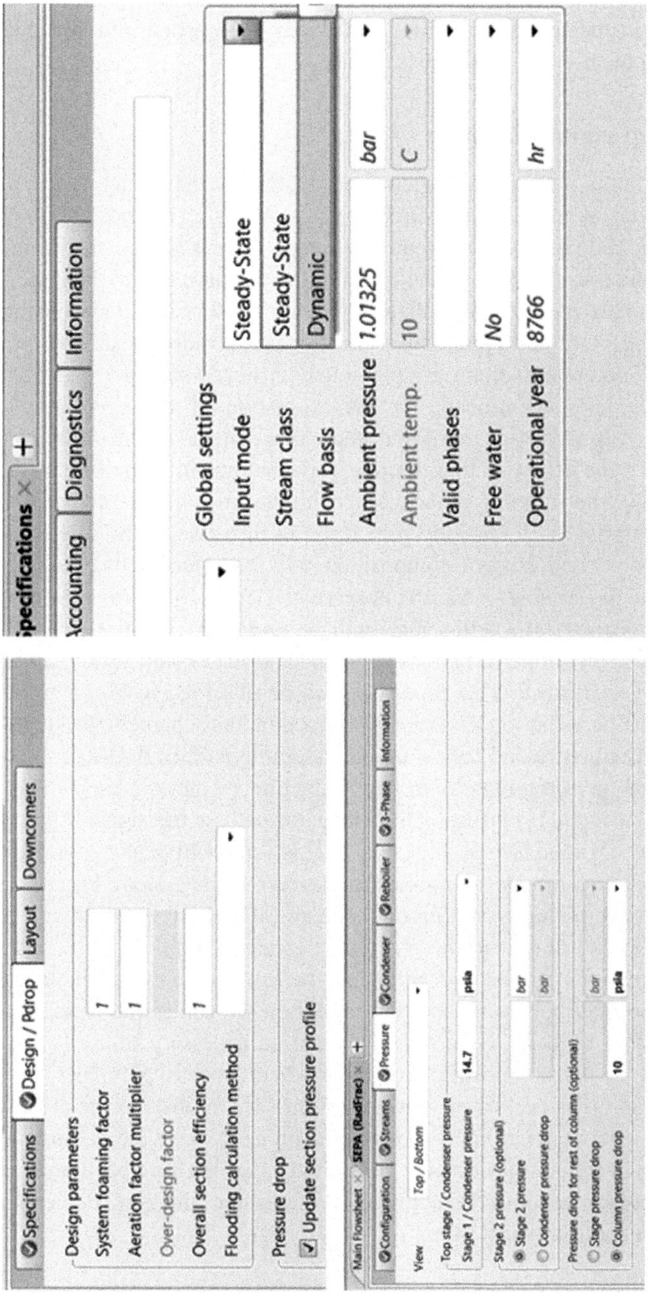

FIGURE 5.14 Preliminary instructions for exporting Aspen Plus steady-state simulations to dynamic state.

and finally select the "Flow Driven Export" option. If there is no major error associated with the hydraulic calculation, a new file will be automatically generated in the folder containing the steady-state aspen plus file. Fig. 5.15 shows graphically the above instructions.

Open-loop control analysis

Now, within the Aspen Plus Dynamic environment, the model that was previously worked on in steady state already exists. Although it is possible to observe the same module that was worked on in steady state, it is necessary to initialize all the values associated with the material balance and energy balance. In this sense, as the first step, the "Initialization" option must be selected from the menu shown in Fig. 5.16. This operation, as mentioned, will allow observing the values in column and currents that were previously in steady state.

As previously mentioned, for the evaluation of the open-loop control properties, it is necessary to make a step change in one of the design variables and observe the effect of this change (perturbation) in some output variable, for example, the mass or molar compositions. In the case of the open-loop control analysis, as closed-loop performed in this case study, it is considered that the dome and bottom composition will be affected by the physically closest design variable. Thus, to observe the effect of perturbations on the dome composition, the reflux ratio will be perturbed. On the other hand, to observe the effect of perturbations on the background composition, the thermal load will be perturbed. This procedure can be observed in Fig. 5.16.

Once the perturbation has been performed on the scheme to be studied, it is necessary to observe and follow up the natural dynamics of the process in the face of the perturbation. With that intention, a new "shape" should be generated, using the corresponding icon located in the upper right part of Aspen Plus Dynamics (see Fig. 5.17). This icon will allow monitoring the change over time of some variable or parameter of the system. Fig. 5.17 shows how by double clicking on a product stream, it is possible to select a product composition and drag it to the previously created shape.

The intention with this procedure is to perturb all the manipulable variables (reflux ratio, thermal loads) and their effect on the controlled variables. Thus, when a reflux ratio R1 is perturbed, it will be necessary to verify the effect caused on all the compositions of the studied process. Similarly, when the thermal load Q1 is perturbed, it will be necessary to observe the effect of this perturbation on all the compositions of the process. Once the perturbation and the change in the compositions or controlled variables have been recorded, the dynamic responses must be adjusted to a transfer function. For example, a function of the following form:

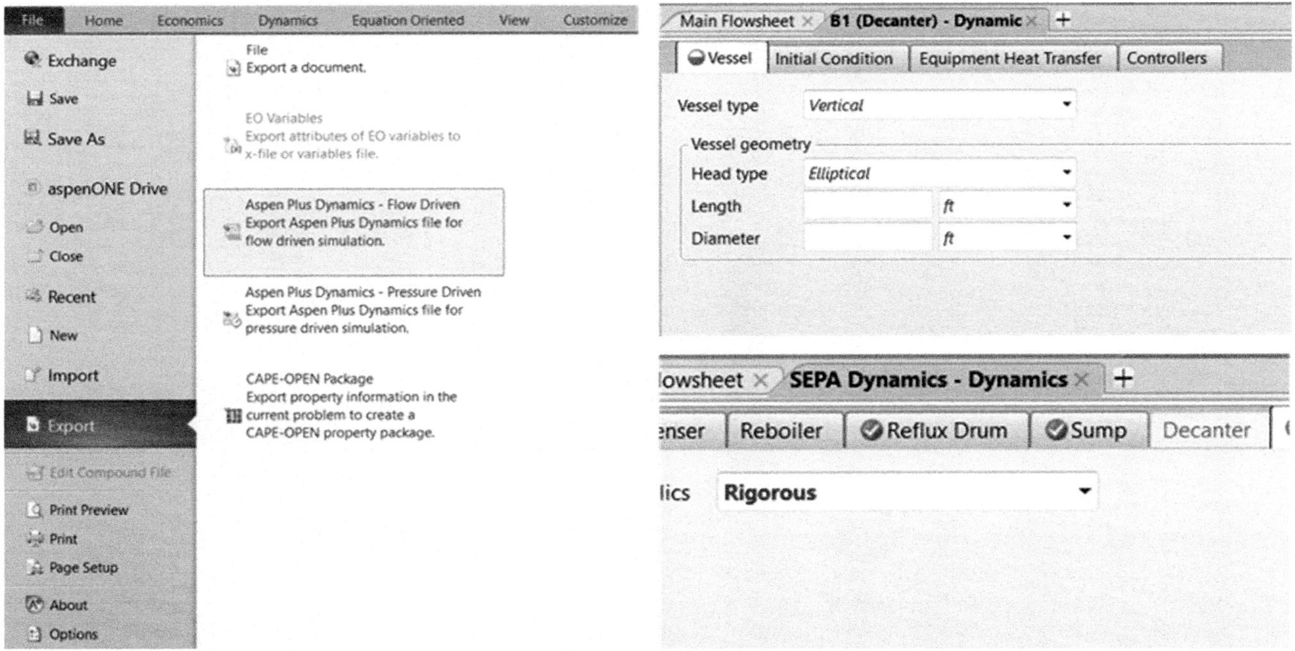

FIGURE 5.15 Preliminary instructions for sizing of the necessary repents in the modeling of column dynamics.

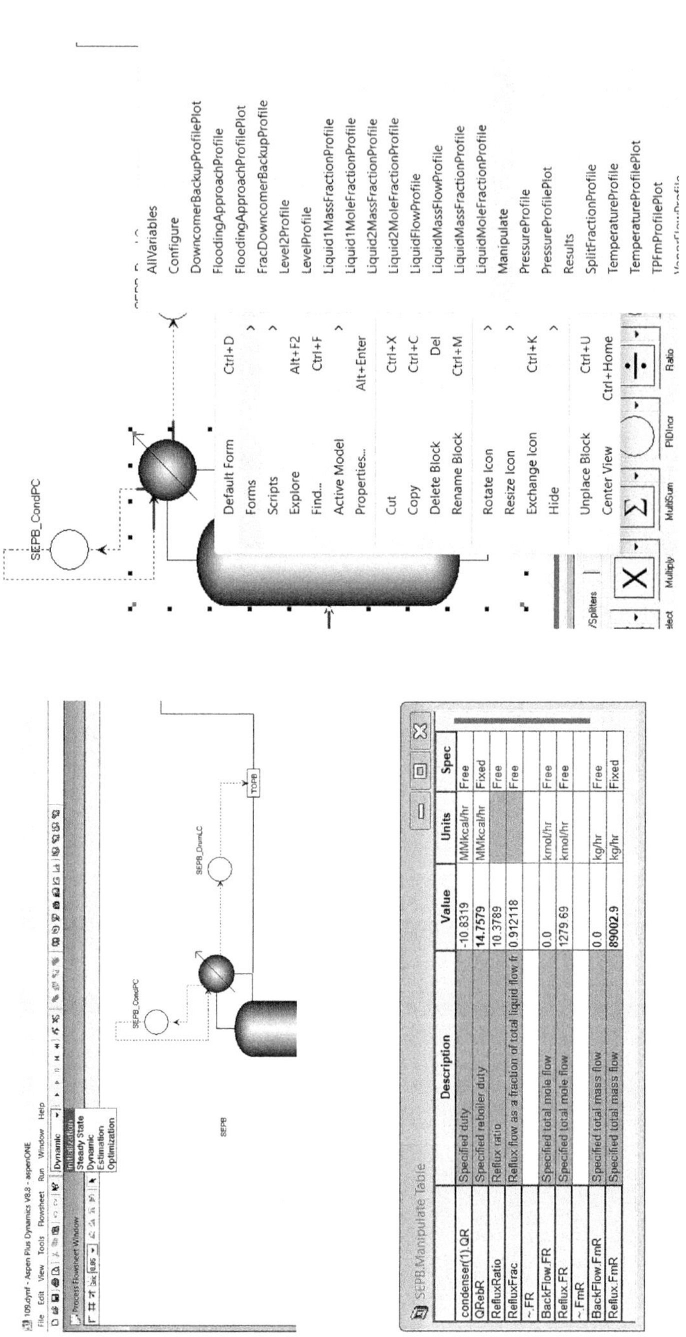

FIGURE 5.16 Preliminary instructions for initializing and perturbing the studied scheme.

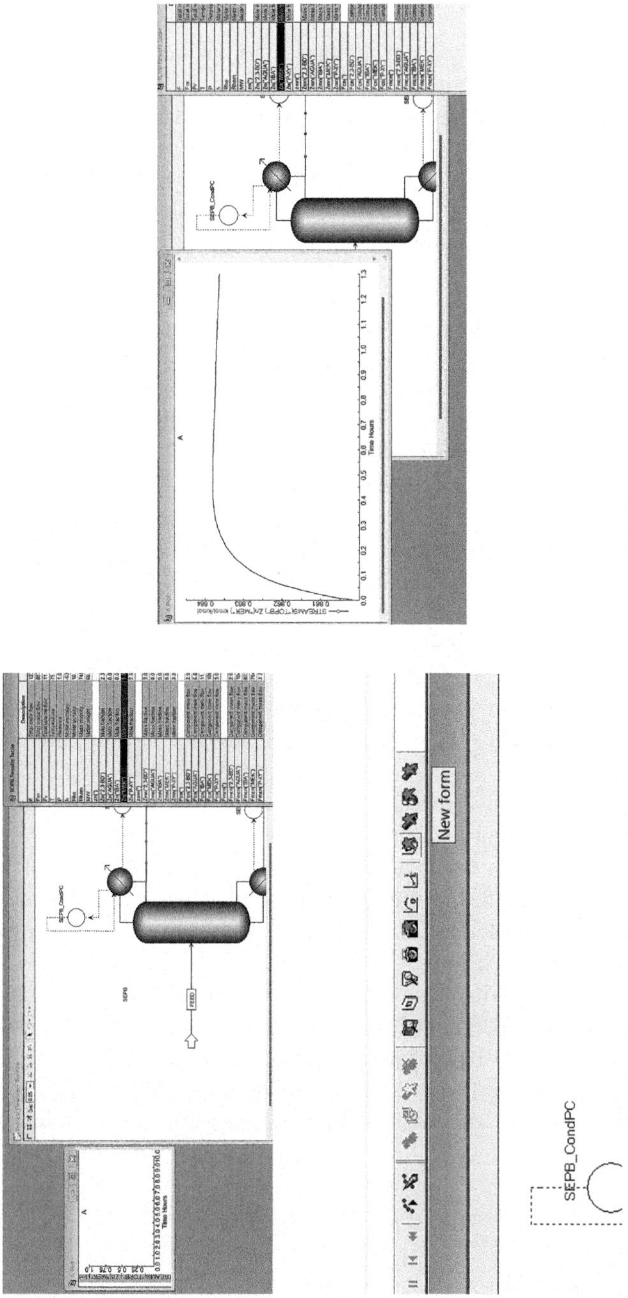

FIGURE 5.17 Preliminary instructions to create a form, and observe the variation over time of some variable, or parameter.

$$g_{i,j} = \frac{k_{i,j}}{\tau_{i,j} \quad s + 1} \tag{5.13}$$

It is possible to use some tool for the fitting, for example Microsoft Excel. Thus, the Aspen Plus Dynamics data is adjusted to a transfer function depending on the degree of the function, the time constant and the damping factor until the error between the theoretical function and the response obtained in the simulator is minimized. With the transfer functions obtained, the transfer matrix G(s) is formed, which relates the compositional changes in the products to the changes in the manipulated variables.

$$\begin{bmatrix} x_A \\ x_B \\ x_C \end{bmatrix} = \begin{bmatrix} g_{11}(s) & g_{12}(s) & g_{13}(s) \\ g_{21}(s) & g_{22}(s) & g_{23}(s) \\ g_{31}(s) & g_{32}(s) & g_{33}(s) \end{bmatrix} \begin{bmatrix} R_1 \\ Q_2 \\ R_3 \end{bmatrix} \tag{5.14}$$

Once the matrix G(s) has been obtained, it is necessary to go from the Laplace domain to the frequency domain by applying the following change $s = \omega j$. For each frequency of interest, the Singular Value Decomposition must be determined, thus obtaining all the singular values. In particular, the condition number (γ^*) is a measure of interest obtained by dividing the maximum singular value by the minimum singular value. The evaluation of the SVD makes it possible to qualitatively determine the controllability of the sequences (evaluation of the control effort), the sensitivity to uncertainties in inputs and outputs, sensitivity to modeling errors and disturbances.

As proposed in the dynamic analysis, the results obtained by means of the DVS depend on the units used for the variables involved, originating a dependence with respect to the magnitude of the changes made in the manipulated variables, since the static gains associated with the linearized model used are dependent on the magnitude of such change, given the nonlinearity of the system under analysis. To obtain correct values of static gains, the magnitude of the change of the manipulated variables must be reduced to about 0.5% of their nominal value to obtain relatively constant gains (Luyben, 1992). Such a size should be even smaller for columns with a higher degree of nonlinearity. This requirement for small changes means that gains obtained from experimental tests would be inaccurate, because for experimental systems 5%–10% changes in their inputs are generally required to eliminate noise in the responses and static gains cannot be determined with great precision. Therefore, the results obtained for experimental columns may not be significant.

Study of closed-loop control properties in a simulator environment

Starting from the same file in dynamic state within the Aspen Plus Dynamics interface, it is necessary to initialize the values of each schematic using the option. For the closed loop study, it is necessary to use a controller. In this exercise a Proportional Integral (PI) controller will be used due to its wide use in industry.

In the object menu at the bottom of the simulation, there is the PIDIncr option, which allows you to select a controller and place it on the workbench next to the schematic to be analyzed. The controller must be connected via "ControlSignal" located in the same menu. When connecting the controller, it must be connected to the current of which you want to control some characteristic (Composition for example). By double clicking on the controller already connected and selecting the "Configure" option, it will be possible to indicate to the simulator whether there will be a new setpoint for the previously selected composition, the parameters with which the controller will work, as well as the type of action of the controller, direct or inverse (see Fig. 5.18).

It is also necessary to add to the workspace a numerical indicator to evaluate the performance of the scheme to be analyzed. For this example we will use the IAE indicator to evaluate the deviation between a new setpoint assigned and the behavior of the controlled variable. In the same way as the controller was added, it is possible to do it with the IAE icon. By double clicking on it, it is possible to monitor the AEI value as well as the input received numerically from the component to be analyzed. By right clicking on the configure option, the new setpoint with which the system will be evaluated must be entered. Once the setpoint of the system to be evaluated has been established, it is possible to quantitatively evaluate the dynamics of the process under a disturbance and under the action of a controller. In order to make a fair comparison between different schemes that are designed to perform the same task, it is necessary to ensure that the best parameters are being used in the controller. In this sense, it is necessary to tune the controller parameters; depending on whether it is a proportional, proportional integral, or proportional integral derivative controller, the parameters to be tuned will depend on the controller. To perform this task, click on the controller panel in the configuration section. Within that section it will be possible to set the working parameters of the controller, as well as the type of controller action (direct or inverse) (see Fig. 5.19).

Once the new setpoint is established in the system, it is possible to follow the dynamics of the process under the action of the controller. This action is possible by inserting an object, and dragging the data of a current, purity for example. Since this action will be performed multiple times, it will be necessary to go back to previous simulation points. To do this, it is possible to use the Snapshot icon at the top right, so it will be possible to go back to all the moments in which an "initialization" process has been performed. Once you have the dynamic response of the process, it is possible to extract the specific data of each moment, by right clicking on the curve in this dynamic, and selecting the option "shows as a table." This option allows you to copy and

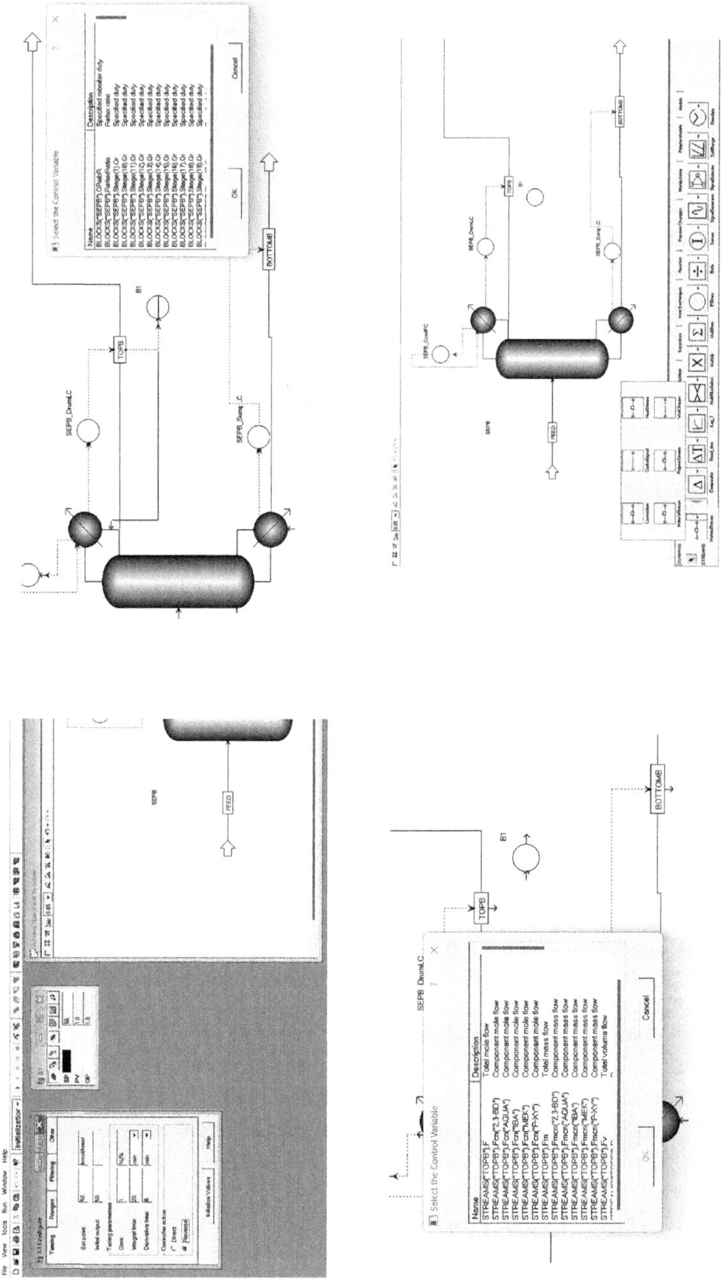

FIGURE 5.18 Preliminary instructions for initializing and connecting a controller.

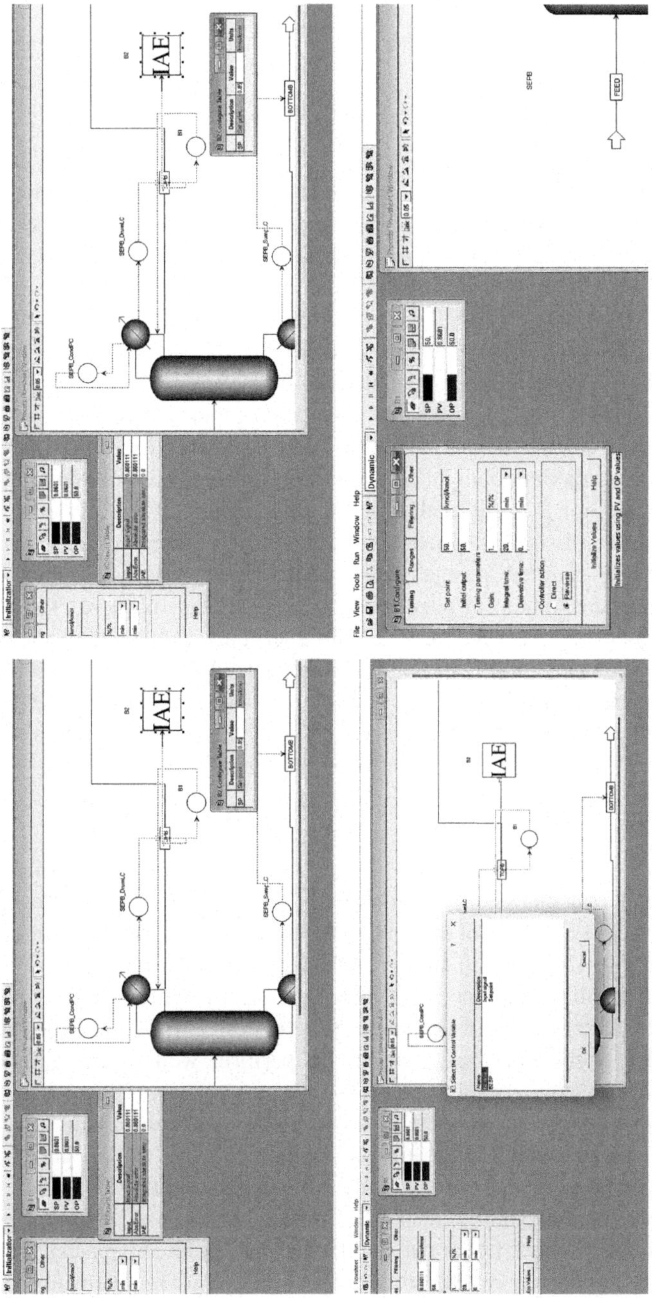

FIGURE 5.19 Preliminary instructions for setting a new setpoint in the controller and performance indicator.

transport the data to any external software. See Fig. 5.20 for instructions on how to do this graphically.

Case study results

Open-loop analysis

The open-loop analysis focused on understanding the natural dynamic behavior of the hybrid distillation processes for biobutanol purification without the application of feedback control. The study revealed that thermally coupled configurations demonstrated superior inherent controllability compared to the conventional configurations. This was evident from the singular value analysis, where the thermally coupled designs exhibited a more compact range of singular values, reflecting reduced sensitivity to disturbances and improved robustness. The condition numbers, derived from the ratio of the largest to the smallest singular value, were notably lower for the thermally coupled schemes. This indicates that these configurations are less susceptible to variations in process inputs, making them inherently more stable and easier to control under dynamic conditions. Additionally, the thermodynamically equivalent configurations exhibited similar benefits, with condition numbers close to those of the thermally coupled designs, showcasing their potential for effective disturbance management in open-loop scenarios. Please observe Fig. 5.21.

The results highlighted the ability of thermally coupled and thermodynamically equivalent configurations to manage feed disturbances more effectively, ensuring minimal deviations from steady-state operation. These findings underline the importance of enhanced energy integration and mass transfer efficiency in achieving robust performance in open-loop conditions.

Closed-loop analysis

The closed-loop analysis evaluated the performance of the hybrid distillation schemes under feedback control conditions using a Proportional-Integral (PI) control strategy. The control performance was assessed through the Integral of Absolute Error (IAE), which quantifies the cumulative deviation of the process variables from their setpoints over time. Among the configurations, the thermally coupled schemes did not achieve the lowest IAE values. The thermodynamically equivalent configurations do perform well in the closed-loop analysis. These configurations demonstrated an effective balance between control performance and operational flexibility, making them a suitable alternative for industrial applications. This performance can be attributed to the energy efficiency and improved internal coupling of these designs, which contribute to better dynamic responses. In contrast, the conventional configurations showed higher IAE values, reflecting their limited capacity to handle disturbances efficiently under closed-loop conditions. The intensified configurations, while reducing equipment complexity, exhibited slightly higher IAE

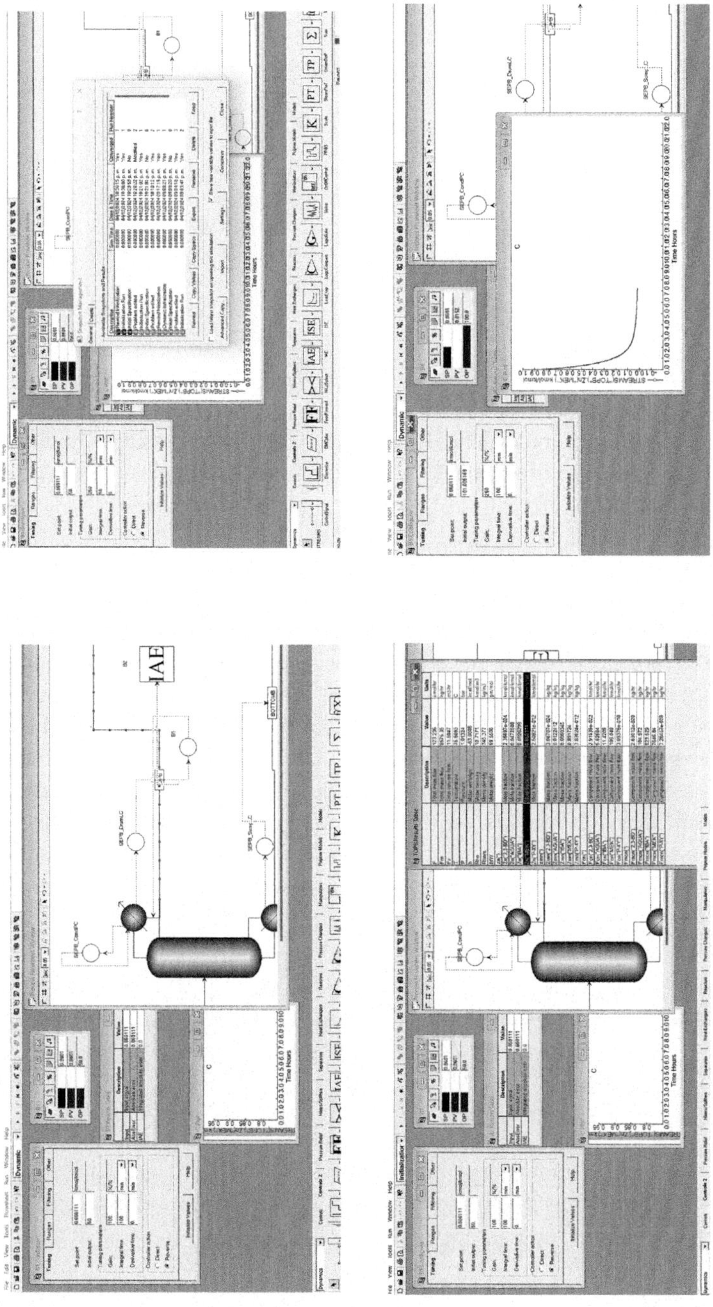

FIGURE 5.20 Instructions for monitoring the dynamic response of the system under analysis.

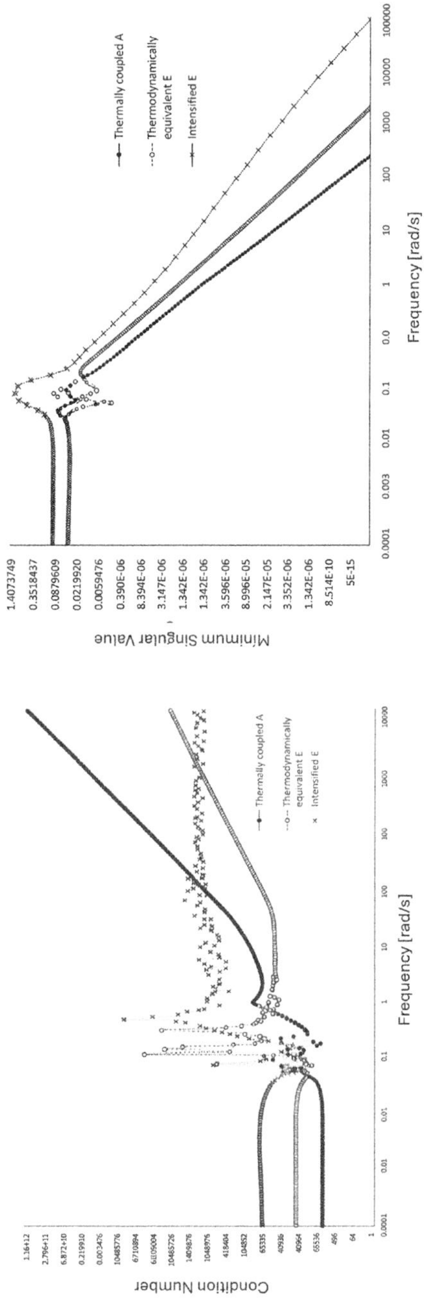

FIGURE 5.21 Minimum singular value and condition number for the best design of each category.

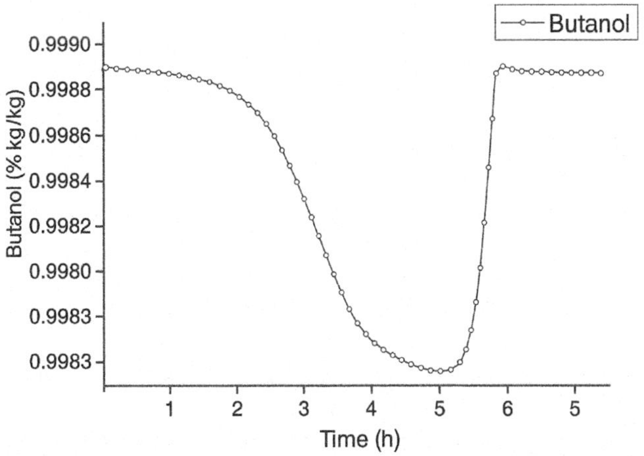

FIGURE 5.22 Closed-loop dynamic response under a feed composition disturbance for butanol outputs in the intensified design.

TABLE 5.7 Kc, τi and IAE values for the tuning process for all study cases.

Set point change	Kc (%/%)	τ_i (min)	IAE
Reference case	250	150	0.04471066
Thermally coupled	250	150	0.02187934
Thermodynamically equivalent	250	20	0.00160158
Intensified	250	140	0.00166647

values than the thermally coupled and thermodynamically equivalent designs due to the reduced number of degrees of freedom for control. In Fig. 5.22, it is possible to observe the intensified alternative under a feed stream disturbance.

The closed-loop analysis underscored the importance of advanced configurations, particularly thermally coupled designs, in achieving robust and sustainable performance under dynamic operating conditions. The evaluation emphasized that these designs not only optimize steady-state metrics, such as energy consumption and cost, but also enhance dynamic controllability, aligning with the goals of sustainable and resilient chemical process operations. Table 5.7 show the Kc and τ_i data, also shows the IAE values.

General conclusions

The findings of this chapter underscore the significant advantages of thermally coupled and intensified configurations in achieving sustainable and efficient separation processes. These designs consistently demonstrated substantial energy savings, reduced environmental impact, and lower total annual costs when compared to conventional alternatives. The dynamic control analyses further emphasized the importance of controllability and operational robustness, with thermally coupled and thermodynamically equivalent configurations showing superior performance under various disturbances. The use of advanced simulation tools, such as Aspen Plus, was instrumental in facilitating the systematic design, optimization, and evaluation of these processes, enabling the integration of thermal coupling strategies and multi-objective optimization. The case studies presented—spanning diphenyl carbonate production, nylon industry effluent recovery, and biobutanol purification—highlight the versatility and effectiveness of these approaches in addressing the economic, environmental, and operational challenges of modern chemical processes. Overall, the results reaffirm the critical role of combining process intensification strategies with simulation-based methodologies to align chemical processes with global sustainability objectives.

References

Alcocer-García, H., Segovia-Hernández, J. G., Prado-Rubio, O. A., Sánchez-Ramírez, E., & Quiroz-Ramírez, J. J. (2019). Multi-objective optimization of intensified processes for the purification of levulinic acid involving economic and environmental objectives. *Chemical Engineering and Processing - Process Intensification, 136*, 123–137. https://doi.org/10.1016/j.cep.2019.01.008

Anastas, P. T., & Warner, J. C. (2000). *Green chemistry: Theory and practice*. https://doi.org/10.1093/OSO/9780198506980.001.0001

Antonakou, E. V., & Achilias, D. S. (2013). Recent advances in polycarbonate recycling: a review of degradation methods and their mechanisms. *Waste and Biomass Valorization, 4*(1), 9–21. https://doi.org/10.1007/s12649-012-9159-x

Aresta, M., Dibenedetto, A., & Angelini, A. (2014). Catalysis for the valorization of exhaust carbon: from CO_2 to chemicals, materials, and fuels. Technological use of CO_2. *Chemical Reviews, 114*(3), 1709–1742. https://doi.org/10.1021/cr4002758

Arvidsson, R., Tillman, A. M., Sandén, B. A., Janssen, M., Nordelöf, A., Kushnir, D., & Molander, S. (2018). Environmental assessment of emerging technologies: recommendations for prospective LCA. *Journal of Industrial Ecology, 22*(6), 1286–1294. https://doi.org/10.1111/jiec.12690

Asbjornsen, O. A. (1985). Chemical process control: An introduction to theory and practice. George Stephanopoulos. *Automatica, 21*(4), 502–504. https://doi.org/10.1016/0005-1098(85)90090-1

Biermann, F., Kanie, N., & Kim, R. E. (2017). Global governance by goal-setting: The novel approach of the UN sustainable development goals. *Current Opinion in Environmental Sustainability, 26–27*, 26–31. https://doi.org/10.1016/j.cosust.2017.01.010

Boddula, R., Shanmugam, P., Srivatsava, R. K., Tabassum, N., Pothu, R., Naik, R., Saran, A., Viswanadham, B., Radwan, A. B., & Al-Qahtani, N. (2023). Catalytic valorisation of biomass-derived levulinic acid to biofuel additive γ-valerolactone: Influence of copper loading on silica support. *Qatar Reactions, 4*(3), 465–477. https://doi.org/10.3390/reactions4030028

Boodhoo, K., & Harvey, A. (2013). *Process intensification for green chemistry: Engineering solutions for sustainable chemical processing*. https://doi.org/10.1002/9781118498521

Bozell, J. J., & Petersen, G. R. (2010). Technology development for the production of biobased products from biorefinery carbohydrates—the US Department of Energy's "top 10" revisited. *Green Chemistry, 12*(4), 539–555. https://doi.org/10.1039/b922014c

Bravo-García, J., Huerta-Rosas, B., Sánchez-Ramírez, E., & Segovia-Hernández, J. G. (2021). Sustainability evaluation of intensified alternatives applied to the recovery of nylon industry effluents. *Process Safety and Environmental Protection, 147*, 505–517. https://doi.org/10.1016/j.psep.2020.11.040

Celińska, E., & Grajek, W. (2009). Biotechnological production of 2,3-butanediol—Current state and prospects. *Biotechnology Advances, 27*(6), 715–725. https://doi.org/10.1016/j.biotechadv.2009.05.002

Contreras-Zarazúa, G., Vázquez-Castillo, J. A., Ramírez-Márquez, C., Pontis, G. A., Segovia-Hernández, J. G., & Alcántara-Ávila, J. R. (2017). Comparison of intensified reactive distillation configurations for the synthesis of diphenyl carbonate. *Energy, 135*, 637–649. https://doi.org/10.1016/j.energy.2017.06.156

Crowl, D. A., & Louvar, J. F. (2011). Chemical process safety: Fundamentals with applications. *Pearson Education, 3*.

Dai, J. Y., Zhao, P., Cheng, X. L., & Xiu, Z. L. (2015). Enhanced production of 2,3-butanediol from sugarcane molasses. *Applied Biochemistry and Biotechnology, 175*(6), 3014–3024. https://doi.org/10.1007/s12010-015-1481-x

Department of Economic and Social Affairs. (2007). *Indicators of sustainable development: Guidelines and methodologies* (Vol 93). Department of Economic and Social Affairs.

Errico, M., Rong, B.-G., Tola, G., & Turunen, I. (2009). A method for systematic synthesis of multicomponent distillation systems with less than N-1 columns. *Chemical Engineering and Processing: Process Intensification, 48*(4), 907–920. https://doi.org/10.1016/j.cep.2008.12.005

Gadalla, M. A., Olujic, Z., Jansens, P. J., Jobson, M., & Smith, R. (2005). Reducing CO_2 emissions and energy consumption of heat-integrated distillation systems. *Environmental Science & Technology, 39*(17), 6860–6870.

Garg, S. K., & Jain, A. (1995). Fermentative production of 2,3-butanediol: A review. *Bioresource Technology, 51*(2–3), 103–109. https://doi.org/10.1016/0960-8524(94)00136-o

Goedkoop, M., & Spriensma, R. (2001). *The eco-indicator 99A damage oriented method for life cycle impact assessment- annex.*

Grossmann, I. E., & Daichendt, M. M. (1996). New trends in optimization-based approaches to process synthesis. *Computers and Chemical Engineering, 20*(6–7), 665–683. https://doi.org/10.1016/0098-1354(95)00201-4

Kiss, A. A., Jobson, M., & Gao, X. (2019). Reactive distillation: Stepping up to the next level of process intensification. *Industrial and Engineering Chemistry Research, 58*(15), 5909–5918. https://doi.org/10.1021/acs.iecr.8b05450

Klema, V., & Laub, A. (1980). The singular value decomposition: Its computation and some applications. *IEEE Transactions on Automatic Control, 25*(2), 164–176. https://doi.org/10.1109/TAC.1980.1102314

Kooi, E. R., Fulmer, E. I., & Underkofler, L. A. (1948). Production of 2,3-butanediol by fermentation of cornstarch. *Industrial and Engineering Chemistry, 40*(8), 1440–1445. https://doi.org/10.1021/ie50464a022

Kumar, A. (1996). *Guidelines for evaluating the characteristics of vapor cloud explosions, flash fires, and bleves. Center for Chemical Process Safety (CCPS) of the AIChE* (1994) (p. 387). New York, NY: Published by the American Institute of Chemical Engineers. ISBN: 0-8169-0474-X.

Luyben, W. L. (1992). *Practical distillation control IEE colloquium (digest)*. https://doi.org/10.1049/ic:20000114

Luyben, W. L. (2013). *Distillation design and control using Aspen™ simulation*. Wiley. https://doi.org/10.1002/9781118510193

Malone, M. F., & Doherty, M. F. (2000). Reactive distillation. *Industrial and Engineering Chemistry Research, 39*(11), 3953–3957. https://doi.org/10.1021/ie000633m

Medina-Herrera, N., Jiménez-Gutiérrez, A., & Mannan, M. S. (2014). Development of inherently safer distillation systems. *Journal of Loss Prevention in the Process Industries, 29*, 225–239.

Moore, C. (1986). Application of singular value decomposition to the design, analysis, and control of industrial processes. In *Proceedings of the American Control Conference* (pp. 643–650). IEEE. https://doi.org/10.23919/acc.1986.4789019

Ogunmakinde, O. E., Egbelakin, T., & Sher, W. (2022). Contributions of the circular economy to the UN sustainable development goals through sustainable construction. *Resources, Conservation and Recycling, 178*, Article 106023. https://doi.org/10.1016/j.resconrec.2021.106023

Penner, D., Redepenning, C., Mitsos, A., & Viell, J. (2017). Conceptual design of methyl ethyl ketone production via 2,3-butanediol for fuels and chemicals. *Industrial and Engineering Chemistry Research, 56*(14), 3947–3957. https://doi.org/10.1021/acs.iecr.6b03678

Sharma, S., & Rangaiah, G. P. (2013). An improved multi-objective differential evolution with a termination criterion for optimizing chemical processes. *Computers and Chemical Engineering, 56*, 155–173. https://doi.org/10.1016/j.compchemeng.2013.05.004

Smith, C. A., & Corripio, A. B. (2005). *Principles and practices of automatic process control*.

Smith, R. (2016). *Chemical process design and integration* (2nd ed., Vol 21).

Song, D., Yoon, Y. G., & Lee, C. J. (2017). Conceptual design for the recovery of 1,3-butadiene and methyl ethyl ketone via a 2,3-butanediol-dehydration process. *Chemical Engineering Research and Design, 123*, 268–276. https://doi.org/10.1016/j.cherd.2017.05.019

Song, C. W., Park, J. M., Chung, S. C., Lee, S. Y., & Song, H. (2019). Microbial production of 2,3-butanediol for industrial applications. *Journal of Industrial Microbiology and Biotechnology, 46*(11), 1583–1601. https://doi.org/10.1007/s10295-019-02231-0

Stephanopoulos, G. (1984). *Chemical process control: An introduction to theory and practice*. Prentice-Hall.

Sundmacher, K., & Kienle, A. (2003). *Reactive distillation: Status and future directions*. Wiley. https://doi.org/10.1002/3527600523

Wu, K. J., Saratale, G. D., Lo, Y. C., Chen, W. M., Tseng, Z. J., Chang, M. C., Tsai, B. C., Su, A., & Chang, J. S. (2008). Simultaneous production of 2,3-butanediol, ethanol and hydrogen with a *Klebsiella* sp. strain isolated from sewage sludge. *Bioresource Technology, 99*(17), 7966–7970. https://doi.org/10.1016/j.biortech.2008.03.062

Zaidi, A., Gainer, J. L., & Carta, G. (1995). Fatty acid esterification using nylon-immobilized lipase. *Biotechnology and Bioengineering, 48*(6), 601–605. https://doi.org/10.1002/bit.260480607

Zhou, H., Cai, Y., & You, F. (2018). Systems design, modeling, and thermoeconomic analysis of azeotropic distillation processes for organic waste treatment and recovery in nylon plants. *Industrial and Engineering Chemistry Research, 57*(30), 9994–10010. https://doi.org/10.1021/acs.iecr.8b00275

Ziegler-nichols, U. (2012). *Controller parameter tuning with aspen HYSYS® V8*.

Chapter 6

Complete thermal coupling: Petlyuk columns and their implementation as divided wall columns

Motivation, background, and brief history of columns with full thermal coupling: Petlyuk columns

Distillation is one of the most indispensable separation processes in the chemical industry, responsible for purifying a wide range of chemicals. However, its widespread use comes with a significant downside: energy consumption. This has positioned distillation as a key contributor to the operational costs and environmental impact of industrial processes. Addressing these energy demands has driven the development of innovative solutions, including thermally coupled systems such as the Petlyuk column. The chemical industry consumes approximately 10%–15% of the total global energy demand, making it one of the largest industrial energy users (Canton, 2021, p. 684–686). Within the chemical industry, separation processes account for 40%–70% of this energy consumption, with distillation alone representing up to 50% of the total energy used (Sholl & Lively, 2016) (see Fig. 6.1). To put this into perspective, distillation columns worldwide are estimated to consume about 3% of the total global energy production (Seider et al., 1998).

Over the years, these energy demands have only grown due to the increasing complexity of industrial processes and the production of more diverse chemical products. Between 1980 and 2020, the energy required for industrial separations increased by an estimated 25%, largely driven by expanding industrial output and more stringent product purity requirements (Rév et al., 2001). This escalation highlights the pressing need for energy-efficient solutions. The concept of thermally coupled distillation addresses these challenges by eliminating redundant reboilers and condensers, which are primary sources of energy waste in conventional column configurations. Thermally coupled systems, such as the Petlyuk column, leverage energy

Thermally Coupled Distillation Columns. https://doi.org/10.1016/B978-0-443-33114-5.00007-7

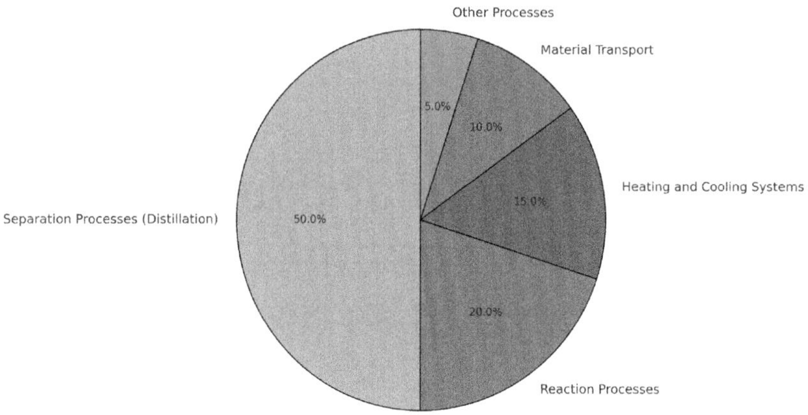

FIGURE 6.1 Energy consumption distribution in the chemical industry.

integration to minimize heat losses, offering the potential to reduce energy consumption by 30%–60% compared to conventional distillation arrangements (Petlyuk, 1965). These savings not only translate to lower operational costs but also align with global efforts to reduce greenhouse gas emissions and improve industrial sustainability.

Historical development of Petlyuk columns

The idea of thermally coupled distillation predates the formal introduction of the Petlyuk column. Early work in the 1940s and 1950s explored the thermodynamic efficiency of distillation systems, laying the groundwork for more advanced designs. However, it was not until 1965 that F.B. Petlyuk, V.M. Platonov, and D.M. Slavinskii formally introduced the concept of a thermodynamically optimal distillation system with full thermal coupling (Petlyuk, 1965). Their seminal paper provided a detailed theoretical framework, demonstrating that such systems could achieve energy efficiency close to the minimum thermodynamic work required for separation. This theoretical innovation spurred further research into the practical implementation of thermally coupled systems. In the 1970s, advancements in computational methods allowed researchers to simulate the performance of these systems, confirming their potential energy savings under various operating conditions. These studies also revealed the intricate balance required to control internal vapor and liquid flows, a challenge that initially limited the adoption of the Petlyuk column (Dünnebier & Pantelides, 1999).

By the 1980s, experimental studies began validating the theoretical predictions. Pilot-scale implementations demonstrated the feasibility of thermally coupled systems, with energy savings of 40%–50% compared to conventional distillation configurations (Triantafyllou & Smith, 1992). These findings provided the impetus for further development, leading to the introduction of the dividing wall column (DWC) in the 1990s. The DWC simplified the design and operation of thermally coupled systems by incorporating the necessary thermal coupling within a single column structure.

The 21st century has seen significant advancements in the understanding and application of Petlyuk and DWCs. Modern computational tools, such as process simulators and advanced control systems, have addressed many of the operational challenges associated with these systems. Today, they are widely recognized as one of the most energy-efficient solutions for multicomponent separations, representing a milestone in the quest for sustainable industrial practices.

This chapter aims to provide a comprehensive overview of thermally coupled distillation systems, focusing on their evolution, energy efficiency, and industrial applicability. Beginning with the motivation and historical development of Petlyuk columns, the discussion extends to their practical implementation as DWCs, energy analysis, and current industrial use. Through detailed case studies, the chapter highlights applications in the purification of complex mixtures and the production of high-value compounds, as well as insights into pilot-scale implementations. Ultimately, this chapter seeks to illustrate the transformative potential of these innovative distillation technologies for modern industrial processes.

Practical implementation of full thermal coupling columns: dividing wall columns

The practical implementation of a Petlyuk column as a DWC is an elegant engineering solution to the complexities of fully thermally coupled distillation systems. While the Petlyuk column offers a theoretical model for energy-efficient separations, its physical realization often involves operational and design challenges. The DWC addresses these issues by consolidating the configuration into a single shell with an internal dividing wall, optimizing energy integration while simplifying the design.

Physical topology of dividing wall columns

The DWC is characterized by a vertical partition that divides a portion of the column into two parallel sections, creating distinct pathways for vapor and liquid flows. This wall typically spans the middle section of the column, encompassing the feed tray region and portions of the rectification and stripping sections. Its placement is designed to ensure effective separation of components while maintaining the thermal coupling required for efficient operation (Górak & Olujić, 2014, p. 1–351).

Inside the DWC, the trays or packing play a crucial role in facilitating vapor–liquid contact. Trays are arranged to distribute liquid evenly across the column, while structured packing, often used in modern DWCs, minimizes pressure drops and enhances mass transfer efficiency (Triantafyllou & Smith, 1992). The vapor generated in the reboiler rises through the column, splitting into two streams as it reaches the dividing wall. Similarly, the liquid descending from the condenser is divided into two streams, directed separately into each section of the column. The dividing wall prevents these streams from mixing, ensuring that each section performs a specific separation task (Dejanović et al., 2010). The reboiler serves as the heat source for the entire column, driving the vaporization necessary for separation. The energy input from the reboiler is optimized through the internal thermal coupling facilitated by the dividing wall. This configuration also helps regulate the reflux ratio across the two sections, maintaining efficient separation while minimizing energy consumption (Schultz et al., 2002) (see Fig. 6.2). The placement of feed and product streams is another critical aspect, as they are introduced and withdrawn at specific trays to maximize separation efficiency.

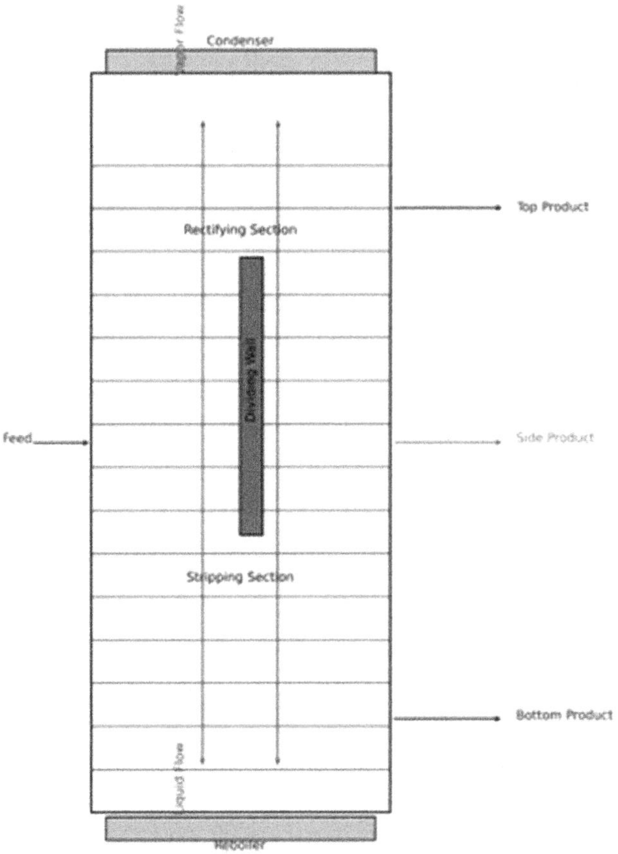

FIGURE 6.2 Schematic of a dividing wall column (DWC).

The physical transformation of a Petlyuk column into a DWC involves significant design considerations. Theoretical models and simulations are used to determine the optimal placement of the dividing wall, feed points, and product withdrawal locations. Advanced tools like Aspen Plus aid in this process, ensuring that the design achieves the desired separation performance while maintaining operational feasibility (Triantafyllou & Smith, 1993, p. 351–360).

The goal of implementing a Petlyuk column as a DWC is to combine the energy-saving benefits of thermally coupled systems with the simplicity of a single-shell design. The DWC consolidates multiple interconnected columns into a single unit, significantly reducing the footprint and capital costs associated with traditional setups. By leveraging the dividing wall to maintain separate vapor and liquid streams, the DWC ensures the thermal coupling necessary for optimal energy efficiency, achieving energy savings of up to 30%–60% compared to conventional configurations (Górak & Olujić, 2014, p. 1–351). In addition to energy efficiency, the DWC design offers advantages in maintenance and operational flexibility. The elimination of external piping and separate reboilers simplifies the mechanical structure, reducing the number of components requiring regular maintenance. Packed DWCs, in particular, benefit from structured packing, which is less prone to fouling and easier to replace compared to trays (Dejanović et al., 2010). This design also facilitates cleaning and inspection, streamlining maintenance procedures and reducing downtime.

Operational control is another advantage of DWCs. With all separation tasks occurring within a single column, the need for complex balancing of flow rates and temperatures across multiple units is eliminated. This integrated approach not only improves process stability but also enhances overall operational efficiency. By addressing both energy and practical challenges, DWCs represent a powerful solution for modern separation processes (Schultz et al., 2002).

Energy analysis of dividing wall columns

DWCs are a groundbreaking advancement in distillation technology, addressing the critical need to reduce the energy demands of industrial separation processes. Distillation accounts for approximately 50% of the energy used in separation processes across the chemical industry, representing a significant operational cost and environmental burden (Asprion & Kaibel, 2010). This challenge is compounded by the increasing global emphasis on sustainability, where energy efficiency and emissions reduction are paramount. Energy analysis in DWCs evaluates their capability to minimize energy consumption through optimized thermal integration. The consolidation of separation tasks within a single column structure reduces redundancy, allowing DWCs to operate with far greater thermodynamic efficiency compared to traditional setups (Petlyuk, 1965). By enabling precise energy utilization and reducing losses, DWCs align with broader industrial objectives, such as reducing greenhouse gas emissions, improving resource efficiency, and lowering production costs. The insights gained from energy analysis form the foundation for justifying the deployment of DWCs in various industrial applications.

A comprehensive energy analysis of DWCs requires an examination of thermodynamic principles, process simulations, and operational data. Central to their energy efficiency is the concept of thermal coupling. In DWCs, the dividing wall facilitates the precise segregation of vapor and liquid flows into two distinct zones, allowing heat transfer to occur exactly where it is needed. This targeted thermal interaction minimizes entropy generation and enhances the exergy efficiency of the system (Dejanović et al., 2011). Thermodynamic analysis reveals that DWCs operate closer to ideal separation conditions compared to traditional systems. The reduction in irreversibilities within the column contributes to a significant improvement in thermodynamic efficiency. For example, exergy analysis of DWCs in specific petrochemical applications has demonstrated efficiency gains of 10%–15% over conventional distillation sequences (CDS) (Fang et al., 2015). These gains result from the minimization of heat losses and the strategic use of internal energy to drive separations.

Entropy generation, a key indicator of system inefficiencies, is notably lower in DWCs due to their integrated design. The division of the column into distinct sections reduces thermal gradients, which are a primary source of entropy increase in conventional systems. This improvement not only translates to energy savings but also supports the long-term sustainability of industrial processes. Simulation tools like Aspen Plus play a crucial role in modeling the energy performance of DWCs. These tools enable the analysis of parameters such as reflux ratios, feed composition, and operating pressures to optimize column design and operation (Fang et al., 2015). By simulating various scenarios, engineers can identify the configurations that maximize energy efficiency while maintaining product quality.

Typical results and energy savings

DWCs deliver substantial energy savings across a range of industrial applications. In the petrochemical industry, where the separation of light hydrocarbons is common, DWCs have reduced energy consumption by up to 50% compared to conventional column sequences (Ignat & Kiss, 2013). For example, in the separation of propane–propylene mixtures, DWCs have demonstrated energy savings of 40%–50%, alongside a reduction in carbon dioxide emissions by approximately 30% (Fang et al., 2015). This dual benefit of energy efficiency and environmental sustainability underscores their value in industrial applications. In fine chemical production, where high-purity separations are critical, DWCs have achieved energy savings of 35%–45% (Dejanović et al., 2014). These savings are particularly significant in processes involving thermally sensitive components, where precise temperature control is essential. In addition, the reduced energy consumption of DWCs contributes to lower operational costs, enhancing their economic feasibility. The

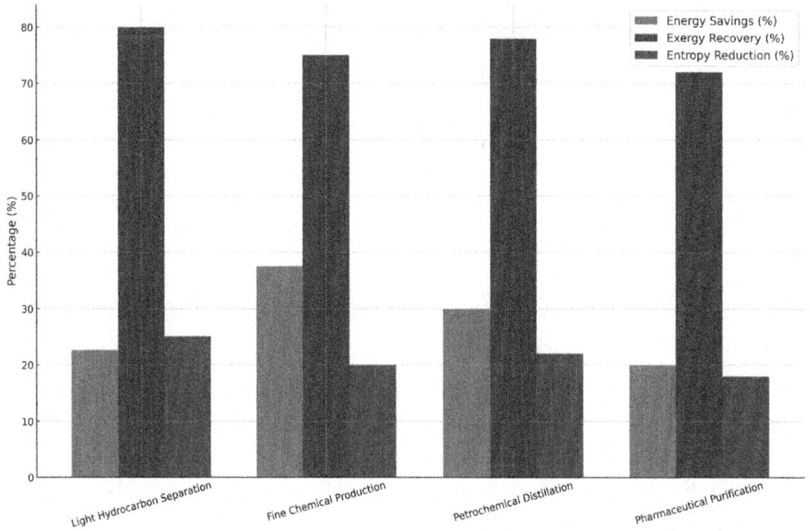

FIGURE 6.3 Performance metrics of dividing wall columns (DWC) in some industries.

thermodynamic efficiency of DWCs is further supported by their ability to achieve separations with lower entropy generation. Studies have shown that entropy production in DWCs can be reduced by 25%–30% compared to conventional systems. This efficiency extends to the exergy balance, with DWCs recovering a higher proportion of the input energy as useful work. In specific applications, exergy efficiency has increased from 65% in traditional columns to 80% in DWCs, representing a substantial improvement in energy performance (Suphanit et al., 2007). Fig. 6.3 shows some typical savings end representative industries.

Limitations and considerations

While DWCs offer unparalleled energy efficiency, their performance is not universally optimal. The energy savings achieved are highly dependent on feed composition and separation complexity. For mixtures with narrow boiling point differences, the potential for energy savings may be diminished due to the additional energy required to achieve sharp separations. Moreover, the thermodynamic advantages of DWCs are most pronounced in systems with well-defined separation tasks; processes with variable feed compositions or fluctuating operating conditions may pose challenges to maintaining optimal energy efficiency. The complexity of thermal integration in DWCs also presents challenges in process control. Maintaining the precise balance of vapor and liquid flows across the dividing wall requires advanced control strategies, particularly in dynamic operational environments. These control requirements, while manageable, add to the initial investment and operational oversight needed for DWCs.

The capital cost of DWCs, though offset by long-term operational savings, can be a barrier to adoption. Advanced simulation and optimization tools are necessary for their design, requiring expertise and resources that may not be readily available in all industrial contexts. Nonetheless, the energy and environmental benefits of DWCs often justify these investments, especially in large-scale applications.

World overview on the use of dividing wall columns and current industrial applications

DWCs have become a cornerstone of modern distillation technology, offering significant energy and cost savings across diverse industrial applications. Since the first industrial implementation by BASF in 1985, their adoption has grown exponentially, with hundreds of units currently operational worldwide. This growth underscores the versatility and efficiency of DWCs in addressing the challenges of energy-intensive separation processes (Lorenz et al., 2018).

Industrial applications of DWCs

DWCs are employed across multiple industries, demonstrating their adaptability to a variety of separation challenges (see Fig. 6.4):

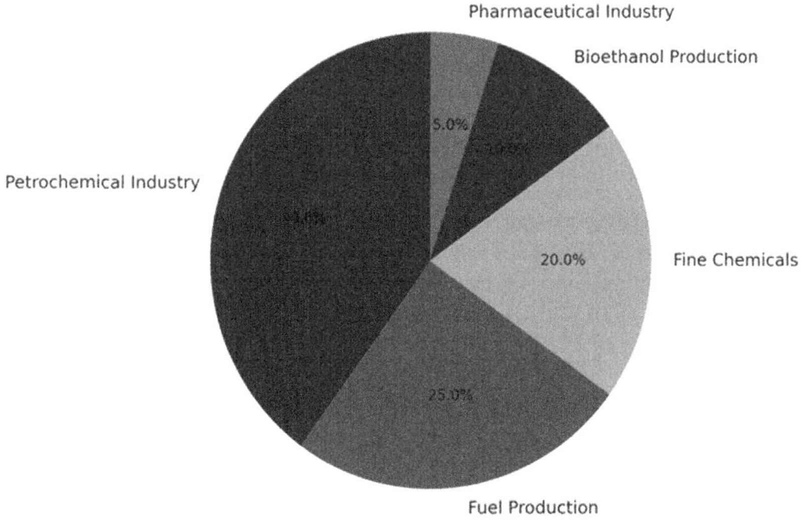

FIGURE 6.4 Industrial applications of DWCs.

- Petrochemical industry: DWCs dominate the petrochemical sector, accounting for approximately 40% of total usage. They are essential for separating complex hydrocarbon mixtures, such as ethylene, propylene, and butadiene, optimizing production processes while reducing energy consumption (Hoyme, 2017, p. 2037–2067).
- Fuel production: Around 25% of DWCs are utilized in refining operations, where they enhance the separation of crude oil fractions. These columns are particularly valuable in increasing the efficiency of producing gasoline, diesel, and other fuels (Hoyme, 2017, p. 2037–2067).
- Fine chemicals industry: DWCs contribute 20% of applications in the fine chemicals sector. They are instrumental in achieving high-purity separations, essential for manufacturing solvents, catalysts, and other specialized chemicals (Yildirim et al., 2011).
- Bioethanol production: DWCs are increasingly used in bioethanol production, representing 10% of their applications. By improving the dehydration process of ethanol, they enhance the efficiency of producing anhydrous bioethanol (Yildirim et al., 2011).
- Pharmaceutical industry: Although representing a smaller share (5%), DWCs are vital in the pharmaceutical sector for purifying active pharmaceutical ingredients. Their ability to maintain stringent quality standards makes them an invaluable tool for drug manufacturing (Kalita & Gentry, 2018).

Global growth and geographical distribution of DWCs

The adoption of DWCs has expanded significantly over the past few decades, with the number of installations increasing from a single unit in 1985 to over 700 units in 2023. This growth reflects a broader trend in the chemical industry toward energy-efficient and sustainable technologies (Hoyme, 2017, p. 2037–2067). The global growth of DWCs is depicted in Fig. 6.5, illustrating the exponential rise in installations over the years. This growth is driven by

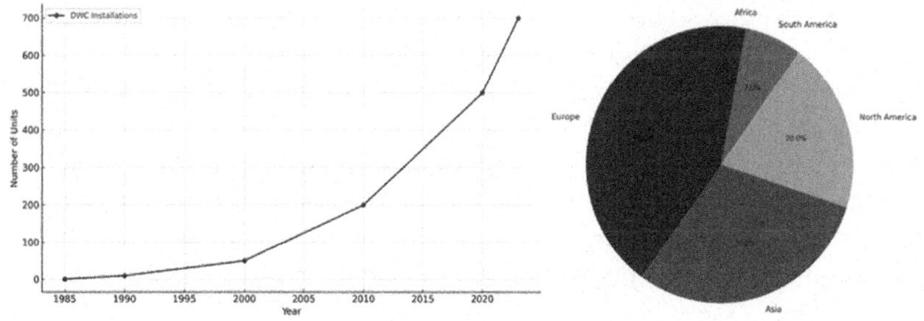

FIGURE 6.5 Global growth and regional adoption of DWCs.

both technological advancements and the rising demand for cost-effective solutions in separation processes.

The adoption of DWCs varies by region, highlighting their widespread applicability:

- Europe: As a leader in adopting DWC technology, Europe accounts for 40% of installations, with a focus on petrochemical and chemical industries (Hoyme, 2017, p. 2037–2067).
- Asia: Rapid industrial growth has positioned Asia as a significant adopter, contributing 30% of DWC installations, particularly in China and India (Bravo, 2019).
- North America: Representing 20% of installations, North America utilizes DWCs extensively in fuel production and chemical manufacturing (Kalita & Gentry, 2018).
- South America and Africa: Together, these regions contribute 10% of installations, reflecting a growing interest in adopting advanced distillation technologies (Yildirim et al., 2011).

Application of DWC-type schemes for the purification of complex mixtures

As discussed in previous sections, DWCs offer numerous advantages over conventional distillation systems, ranging from significant energy savings to enhanced separation efficiency. One of the most remarkable features of DWCs is their ability to handle complex mixtures, including those exhibiting azeotropic behavior or closely boiling components. This capability is largely attributed to the inherent thermodynamic efficiency of DWCs and their advanced design, which facilitates precise control over vapor and liquid flows within the column (Rong, 2011). DWCs achieve superior separation performance by integrating thermal coupling and reducing internal inefficiencies. This integration minimizes the need for additional equipment, such as intermediate reboilers and condensers, while simultaneously reducing entropy generation. The dividing wall creates distinct zones within the column, allowing for the effective separation of multicomponent mixtures in a single unit. This unique feature not only simplifies the process but also reduces operating costs, energy consumption, and environmental impact (Piszczek et al., 2023).

The ability of DWCs to separate complex mixtures makes them particularly suitable for challenging industrial applications. These include mixtures with azeotropic behavior, where conventional distillation methods struggle due to the thermodynamic complexities involved. The design of DWCs enables the manipulation of internal column conditions to overcome such challenges, often resulting in higher purity products and greater process efficiency (Errico et al., 2017).

To illustrate the versatility and effectiveness of DWCs in handling complex separations, this section presents two notable case studies. The first case explores the purification of furfural, a bio-based chemical with significant industrial importance. Its separation from aqueous mixtures poses considerable challenges due to azeotropic behavior and closely boiling components. The application of a DWC for this separation has shown promising results, simplifying and optimizing the process while achieving enhanced separation efficiency.

The second case study focuses on the purification of anisole, a valuable aromatic ether widely used in chemical synthesis. Anisole is often found in mixtures involving azeotropic components, making its separation particularly challenging. The use of a DWC in this context has demonstrated improved energy efficiency, reduced operational complexity, and higher product purity, establishing it as an effective solution for such complex separations.

The following subsections delve into the details of these case studies, showcasing the significant advantages of DWCs in real-world applications.

Case study: Purification of furfural

Problem statement and methodology

Furfural is a renewable chemical derived from lignocellulosic biomass, particularly rich in pentoses such as xylose. Its applications span multiple industries, including the production of resins, solvents, and biofuels. Despite its versatility, furfural's production is thermodynamically challenging due to azeotropic behavior and closely boiling by-products. The conventional Quaker Oats process, while effective, is highly energy-intensive, driving the need for alternative methods that integrate thermal coupling, such as DWCs. This case study explores the production of furfural from wheat straw, a widely available agricultural residue. The study focuses on the separation zone and evaluates DWCs as a promising alternative to conventional configurations (Contreras-Zarazúa et al., 2019).

The process begins with a biomass feed of 10 metric tons per day of wheat straw, obtained at a cost of $25 per ton. In the pretreatment zone, wheat straw undergoes a combination of mechanical and chemical processes to release fermentable sugars. The key steps include acid hydrolysis using 0.5% sulfuric acid at a temperature of 150°C (range may vary between 150 and 220°C) for a duration of 60 minutes, achieving a hemicellulose conversion yield of 80%. The main reactions in this zone are:

$$C_5H_8O_4 + H_2O \rightarrow C_5H_{10}O_5 \tag{6.1}$$

$$C_6H_{10}O_5 + H_2O \rightarrow C_6H_{12}O_6 \tag{6.2}$$

$$C_5H_{10}O_5 \rightarrow C_5H_4O_2 + H_2O \tag{6.3}$$

$$C_5H_{10}O_5 \rightarrow 4.68CH_4O \tag{6.4}$$

$$C_{10}H_{13.9}O_{1.3} \rightarrow C_{10}H_{13.9}O_{1.3} \tag{6.5}$$

Eqs. (6.1) and (6.2) illustrate the breakdown of hemicellulose and cellulose into xylose and glucose, respectively. Eq. (6.3) describes the dehydration of xylose to produce furfural, while Eq. (6.4) represents the conversion of xylose into acetic acid. Eq. (6.5) corresponds to the degradation of xylose resulting in methanol. Lastly, Eq. (6.6) depicts the solubilization of lignin into soluble lignin, which is a physical process. The dehydration of glucose has been excluded from consideration due to its low yield (Contreras-Zarazúa et al., 2022).

The reactive zone converts xylose into furfural through acid-catalyzed dehydration. The reaction is carried out in a packed bed reactor at 190°C and 13.14 atm pressure. The dehydration reaction is as follows:

$$C_5H_{10}O_5 \rightarrow C_5H_4O_2 + 3H_2O \quad (6.6)$$

The furfural yield in this stage is approximately 53% starting from xylose, with by-products such as formic acid and acetic acid further complicating the downstream separation.

The separation zone is the most challenging step due to the azeotropic nature of the furfural-water mixture and the presence of low-boiling impurities. Conventional distillation struggles with these complexities, making DWCs a viable alternative. The DWC integrates multiple separation tasks into a single column, reducing energy consumption and simplifying the process configuration. During separation stage, the main challenges include: Handling azeotropic behavior between furfural and water (see Fig. 6.6); Achieving high purity (>99%) and recovery (>95%) for furfural.

As mentioned, the purification of furfural from biomass-derived streams presents significant challenges due to the thermodynamic complexity of the mixture. The azeotropic behavior between furfural and water, combined with the presence of closely boiling impurities such as acetic acid and formic acid, makes this separation particularly demanding. These challenges necessitate high

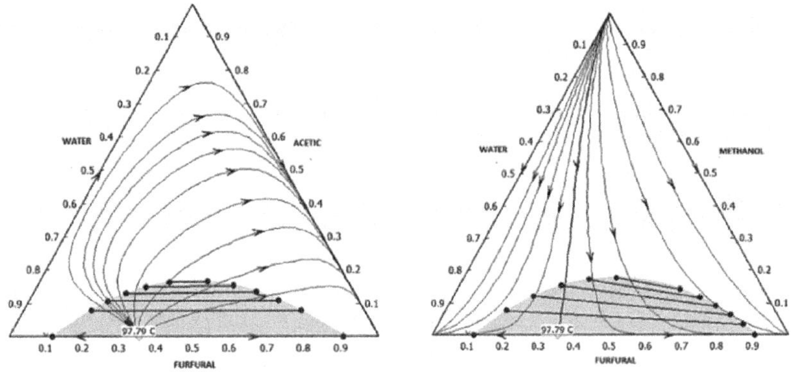

FIGURE 6.6 Ternary diagrams for the mixture water-furfural- methanol- acetic acid.

energy inputs and advanced separation techniques to achieve the desired furfural purity and recovery. Addressing these thermodynamic hurdles has been the driving factor behind the development of both conventional and innovative separation strategies. The conventional Quaker Oats process employs a series of separation units, including three distillation columns and a decantation step. The first column performs pre-fractionation, separating lighter components such as acetic acid and formic acid from the furfural-water mixture. The second column focuses on furfural recovery from water, which is particularly energy-intensive due to azeotropy. Following this, a decanter separates the immiscible phases, isolating furfural-rich streams from water. Finally, a third distillation column is used for final purification, achieving furfural purity levels exceeding 99%. While this method is effective, the extensive equipment and energy requirements highlight the limitations of conventional approaches in handling thermodynamically complex mixtures (see Fig. 6.7).

To overcome these limitations, the integration of a DWC has been proposed as an innovative alternative. The DWC consolidates the functions of multiple distillation columns into a single unit, streamlining the separation process. By incorporating a dividing wall, this column enables simultaneous prefractionation, azeotropic separation, and final purification (see Fig. 6.8). This design minimizes the need for intermediate reboilers and condensers, leading to significant energy savings and reduced operational costs. A decantation step is retained in this configuration to handle immiscible phases effectively, but the overall process complexity is markedly reduced.

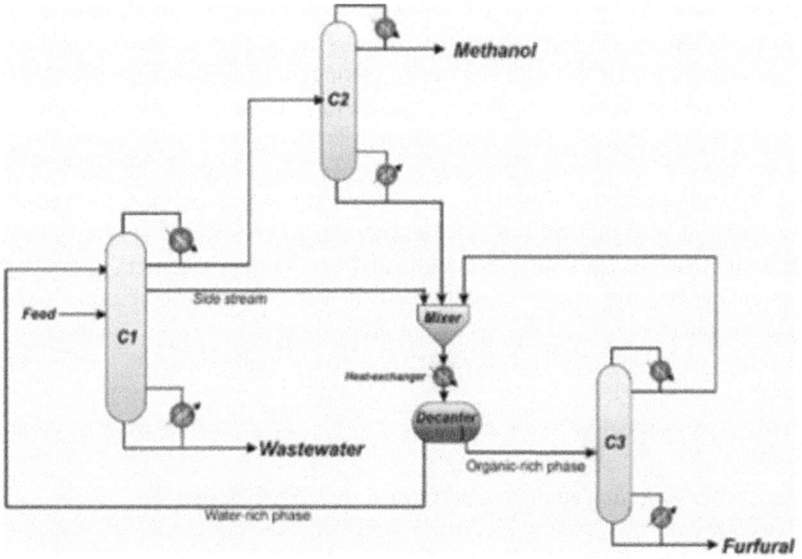

FIGURE 6.7 Quaker oats process to purify furfural.

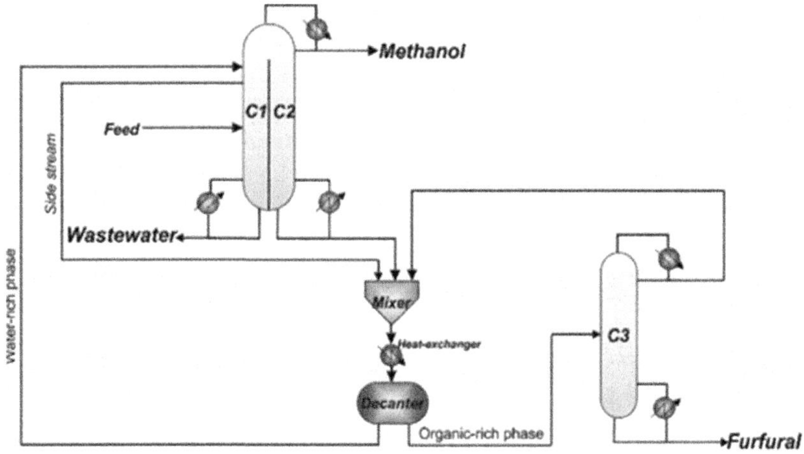

FIGURE 6.8 DWC based process to purify furfural.

Both separation proposals were designed using Aspen Plus. Once both proposals were designed, an optimization process continued. The optimization begins with each simulated in Aspen Plus using data from prior research. During the prescreening stage, alternatives are assessed based on two key metrics: total annual cost (TAC) and the Eco-indicator 99 (EI99), a life cycle analysis tool for environmental impact evaluation. This stage identifies the most promising configuration for each biomass type. The second step focuses on optimizing the selected configurations using a hybrid platform that links Aspen Plus with the differential evolution with tabu list optimization algorithm. This method iteratively refines process parameters to achieve optimal trade-offs between TAC and EI99 while meeting purity (99.2% for furfural and 99% for methanol) and mass flowrate (1000 kg/h for furfural and 200 kg/h for methanol) constraints. The optimization platform, implemented in Visual Basic within Excel, ensures robust performance across multiple iterations. The optimization procedure highlights the adaptability of the proposed approach for different biorefineries and chemical production processes. It underscores the importance of integrating economic and environmental criteria to develop sustainable and cost-effective solutions. This methodology can be extended to other biochemicals, biomass types, and processing technologies, emphasizing its utility in advancing green chemistry and process intensification (Contreras-Zarazúa et al., 2019).

Results

Once process optimization was performed, the DWC design was particularly noteworthy for its ability to reduce remixing phenomena, a primary inefficiency in traditional distillation processes indicates that the DWC reduced energy requirements by approximately 20%–30% compared to the Quaker Oats process. This reduction stems from better distribution of vapor and liquid phases within the column, minimizing the back-mixing of components. In addition to energy savings, the DWC achieved significant reductions in the TAC, with a decrease of 15% compared to conventional designs.

Table 6.1 shows the design parameters for both conventional and DWC configurations. The DWC required fewer trays and a smaller reboiler duty while maintaining high furfural purity (99.2%) and recovery rates. The TCC

TABLE 6.1 Design parameters for both alternatives, conventional and DWC-based.

Design variable	Conventional process	DWC based process
Stages E1	–	–
Stages C1	55	81
Stages C2	12	83
Stages C3	6	11
Feed stage of water-rich phase C1	22	6
Feed tray C1	27	39
Tray of side stream C1	16	33
Feed stage C2	8	–
Feed stage C3	3	4
Diameter of C1 (m)	0.44	–
Diameter of C2 (m)	0.42	2.57
Diameter of C3 (m)	1.63	1.23
Top pressure (atm)	1	1
Reflux ratio of C1	18.5	–
Reflux ratio of C2	0.21	25.14
Reflux ratio of C3	0.208	0.233
Furfural stream (kg/h)	6308.37	6300.8
Entrainer flowrate (kg/h)	–	–
Side stream flowrate (kg/h)	3882	3821.96
Liquid stream flowrate (kg/h)	–	4696
Vapor stream flowrate (kg/h)	–	38,790.3

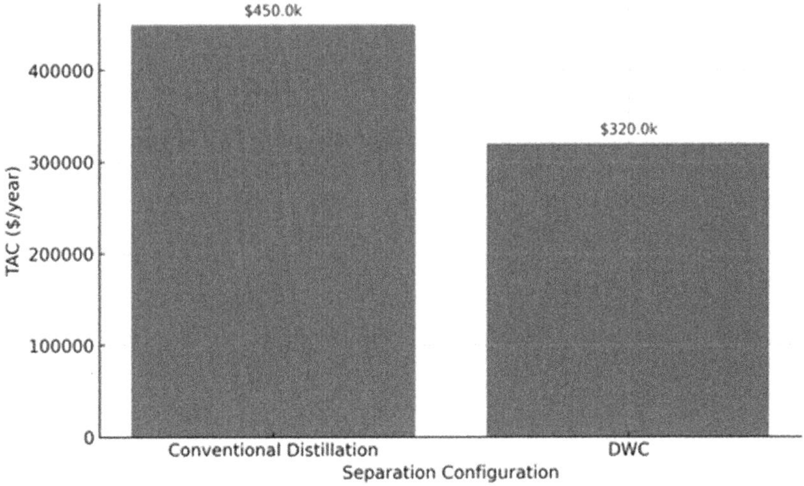

FIGURE 6.9 TAC for conventional and DWC-based separation alternatives.

configuration showed similar energy performance but was marginally less costly due to a smaller column size and simpler integration requirements.

The optimization phase confirmed these findings, with Pareto front analysis demonstrating superior trade-offs between TAC and Eco-Indicator 99 (EI99) for the DWC process. These results underscore the environmental advantages of the DWC, which generates lower emissions and reduces the overall environmental footprint of furfural production. Fig. 6.9 shows TAC results for both alternatives.

Case study: Purification of anisol

Anisole, also known as methoxybenzene, is a versatile aromatic ether widely used in the chemical and pharmaceutical industries. It serves as an intermediate in the production of perfumes, dyes, and pharmaceuticals, as well as a precursor for synthesizing agrochemicals and other specialty chemicals. Given its broad industrial applications, the demand for anisole has steadily increased, underscoring the need for efficient and sustainable production and purification methods (Delolo et al., 2019).

The production of anisole is traditionally achieved through methylation of phenol using methanol or dimethyl sulfate as methylating agents. From a sustainability perspective, incorporating renewable feedstocks such as bio-based phenol or green methanol offers a pathway to reduce the carbon footprint associated with its production (Zhu et al., 2010). Aligning with the principles of green chemistry, the adoption of intensified separation technologies such as thermally coupled configurations and DWCs provides opportunities to enhance the sustainability of anisole production. This section aims to explore optimized separation strategies for anisole, evaluating configurations that integrate thermal coupling and DWCs to achieve improved energy efficiency and environmental performance (Tinoco-Sáenz et al., 2022).

Purification of anisole

The purification of anisole typically involves separating it from a feed stream containing phenol, water, and unreacted methanol. The feed stream has a flowrate of 5000 kg/h and is characterized by a complex thermodynamic behavior, including azeotropic interactions between anisole and water, and phenol and water. These azeotropes complicate the separation process, making it highly energy-intensive and requiring advanced separation techniques (Tinoco-Sáenz et al., 2022).

Several separation alternatives were proposed to address these challenges (Tinoco-Sáenz et al., 2022), including:

- CDS:
  This configuration involves multiple distillation columns operating independently to separate anisole from other components. While straightforward, it demands significant energy input due to the high reflux ratios and reboiler duties required (see Fig. 6.10).

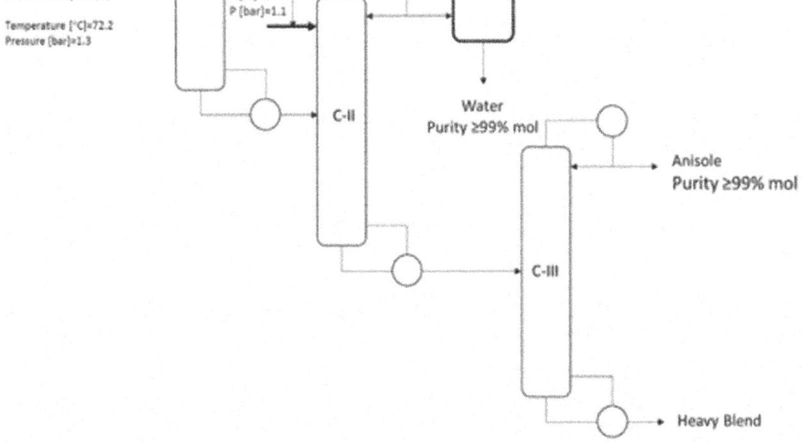

FIGURE 6.10 Conventional alternative for anisole purification.

- Thermally coupled column (TCS1 and TCS2):
  A TCC integrates two columns thermally, eliminating the need for intermediate condensers and reboilers. This configuration reduces energy consumption by redistributing heat within the system but requires careful design and operation to ensure stability (see Figs. 6.11 and 6.12).
- DWC:
  The DWC consolidates separation tasks into a single unit by incorporating a dividing wall, enabling simultaneous separation of anisole, phenol, and water. This design minimizes energy losses due to remixing and reduces the overall footprint of the process.

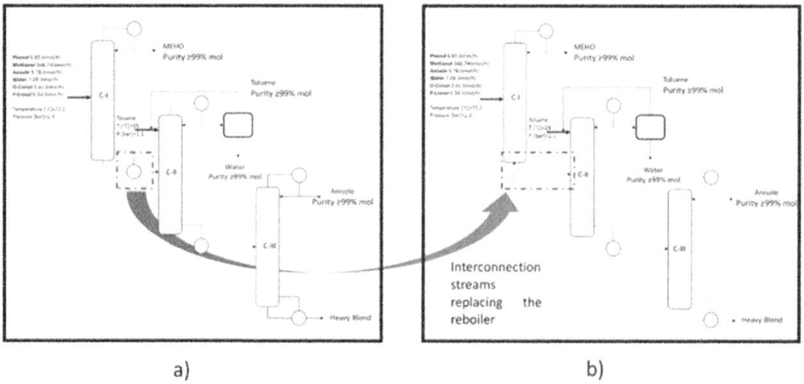

FIGURE 6.11 (a) Diagram of the conventional anisole separation process (CS), (b) diagram of the thermal coupled sequence 1 (TCS1).

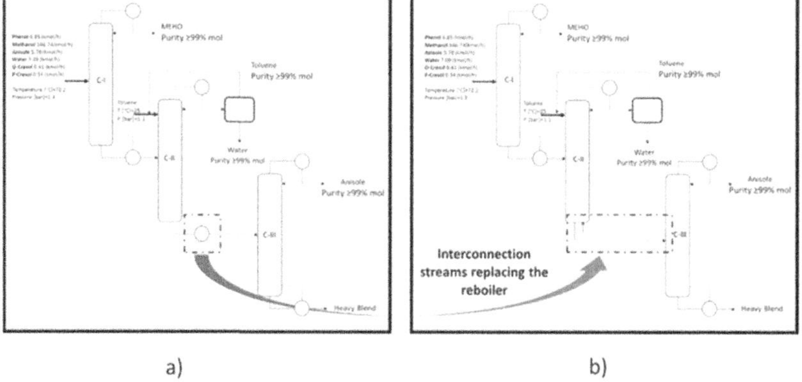

FIGURE 6.12 (a) Diagram of the conventional anisole separation process (CS), (b) diagram of the thermal coupled sequence 2 (TCS2).

Optimization procedure

Optimization was performed using a Differential Evolution algorithm with a Tabu List (DETL) (Srinivas & Rangaiah, 2007), focusing on minimizing the TAC, environmental impact (measured using the Eco-Indicator 99). The objective functions also considered process controllability, measured by the Condition Number (CN), and inherent safety (IR). The objective function considered is expressed as follow:

$$\mathrm{Min(TAC, IR, CN)} = f\left(N_t, N_f, D_f, R_f, S_f, N_s\right)$$

Being the equilibrium stages, feed stages, diameter, reflux ratio, solvent flow, and the solvent feed stage. All variables were systematically optimized to achieve an ideal balance between economic, environmental, and operational goals. Key constraints included maintaining anisole purity at 99.5% and recovery at 98%. Parameters such as reflux ratios, feed locations, and tray counts were systematically optimized to achieve an optimal balance between cost and environmental performance.

Results and discussion

The optimization results highlighted the DWC as the most effective configuration, achieving significant reductions in TAC, energy consumption, and environmental impact compared to other alternatives. Key performance metrics are summarized below in Table 6.2.

The DWC outperformed all other configurations, achieving a 2.43% reduction in TAC, a 3.35% reduction in EI99, and significantly lower energy consumption. Its superior controllability (CN = 356.65) further highlights its practicality for industrial implementation. Compared to the conventional sequence, the DWC offers a holistic improvement in economic and environmental metrics, demonstrating its potential for sustainable anisole purification. Table 6.3 shows the main parameter design for the DWC alternative. In the same way, Fig. 6.13 shows the complete mass balance for the optimal DWC alternative.

TABLE 6.2 Objective functions from optimal designs.

Configuration	TAC ($/year)	EI99	CN	IR (prob/year)
CS	3,250,219	1.489	3748	6.87E-05
TCS1	3,866,473	1.548	2582	3.22E-04
TCS2	3,169,914	1.444	3462	6.91E-05
DWC	3,171,204	1.439	356	6.85E-05

TABLE 6.3 Design parameters for the DWC design.

Variable	DWC
Number of stages C–I	24
Number of stages C-II	21
Number of stages C-III	62
Feed stage C–I	12
Feed stage C-II	16
Toluene feed stage	3
Vapor feed stage	16
Liquid feed stage	37
Side product stage	23
Distillate flow rate C-I (kmol/h)	347.05
Distillate flow rate C-III (kmol/h)	19.59
Lateral product flow (kmol/h)	5.68
Vapor interconnection flow (kmol/h)	31.63
Liquid interconnection flow (kmol/h)	0.48
Toluene flow rate (kmol/h)	12.94
Reboiler duty C–I (cal/s)	1,450,114.81
Reboiler duty C-III (cal/s)	112,224.21
Diameter C–I (meters)	1.37
Diameter C-II (meters)	0.81
Diameter C-III (meters)	0.58
TAC ($/y)	3,096,914.00
CN	29,367.02
IR (1/y)	6.89E-05

Complex reactive DWC schemes for the production of high value-added compounds

DWCs are not only valuable for conventional separations but also demonstrate exceptional potential in complex scenarios that involve simultaneous reaction and separation processes. This integration, often referred to as reactive distillation within a DWC, leverages the synergistic interaction between reaction kinetics and separation dynamics to optimize process efficiency. By

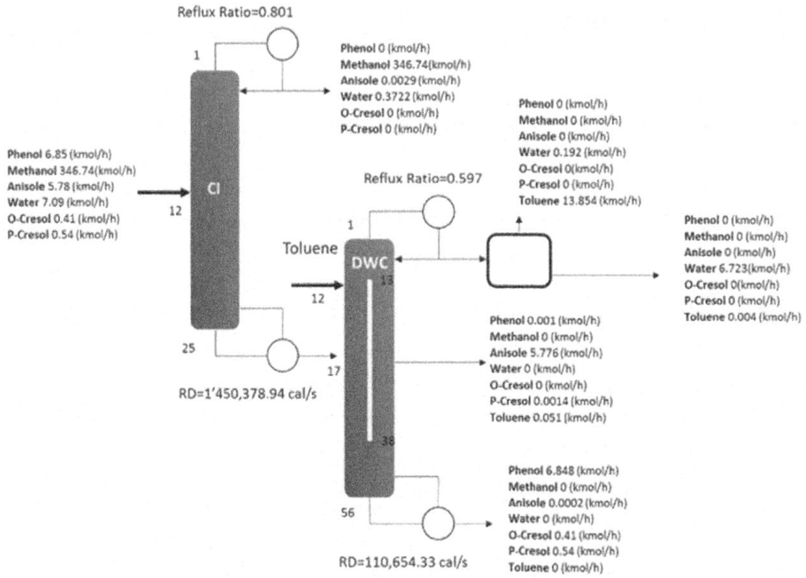

FIGURE 6.13 Optimal DWC sequence design for anisole separation.

combining these operations in a single unit, DWCs can reduce energy consumption, enhance product selectivity, and minimize equipment footprint compared to traditional multi-unit setups (Kiss & Ignat, 2012).

The use of reactive DWCs offers several distinct advantages. First, the direct coupling of reaction and separation shifts the reaction equilibrium, increasing conversion rates and reducing the need for excess reactants. Second, the elimination of intermediate units and reduced recycling loops contribute to significant cost and energy savings. Finally, the compact nature of reactive DWCs supports intensified processes that align with sustainability goals, making them particularly suitable for the production of high value-added compounds (Sundmacher & Kienle, 2003).

This section of the chapter explores two case studies that illustrate the application of reactive DWCs for complex processes. In section Case study: Production of triethyl citrate, the production of triethyl citrate is examined, a compound widely used in the pharmaceutical and food industries as a plasticizer and emulsifier. This process highlights the efficiency of DWCs in handling equilibrium-limited reactions with simultaneous separation of by-products. In section Case study: Production of FAMEs, the focus shifts to the production of fatty acid methyl esters (FAMEs), commonly known as biodiesel, emphasizing the ability of DWCs to streamline transesterification reactions while achieving high product purity and sustainability metrics.

Through these case studies, the versatility and benefits of reactive DWCs in producing high value-added compounds will be comprehensively analyzed.

Case study: Production of triethyl citrate

Triethyl citrate (TEC) is a widely used organic compound known for its versatile applications in various industries. Its primary use as a non-toxic plasticizer makes it essential in pharmaceuticals, where it is utilized in coatings for tablets and capsules, and in the food industry as an emulsifier in beverages and confectioneries. Additionally, TEC is employed in cosmetics and personal care products due to its biodegradability and low environmental impact. As global industries shift toward sustainable practices, optimizing the production of TEC has become a pressing need, especially in minimizing its environmental footprint and energy demands (Zuber et al., 2019). TEC is produced through the esterification of citric acid with ethanol, a reaction that yields intermediate esters such as monoethyl and diethyl citrate before forming the desired triethyl citrate. Traditionally, TEC production relies on batch reactors that often require multiple processing steps, high energy input, and complex post-reaction separations. These factors contribute to inefficiencies in conventional methods, highlighting the need for integrated processes (Finkelstein & Gold, 1959).

From a sustainability perspective, reactive distillation in advanced configurations such as DWCs offers a promising alternative. These systems integrate reaction and separation into a single unit, enhancing energy efficiency and reducing equipment requirements. The integration also aligns with green chemistry principles by minimizing waste generation and improving resource utilization. The objective of this section is to explore TEC production using advanced reactive distillation configurations, with a focus on thermally coupled designs and DWCs (Santaella et al., 2017).

Esterification of citric acid and conceptual design

The esterification of citric acid with ethanol involves a series of equilibrium-limited reactions that form TEC in multiple stages (Kolah et al., 2008):

$$C_6H_8O_7 + C_2H_5OH \rightarrow C_6H_7O_7C_2H_5 + H_2O$$

$$C_6H_7O_7C_2H_5 + C_2H_5OH \rightarrow C_6H_6O_7(C_2H_5)2 + H_2O$$

$$C_6H_7O_7(C_2H_5)2 + C_2H_5OH \rightarrow C_6H_5O_7(C_2H_5)3 + H_2O$$

The kinetics of these reactions indicate that the formation of TEC is favored by high ethanol concentrations, which shift the equilibrium toward the products, and by efficient removal of water to drive the reaction forward. A major challenge in conventional systems is the accumulation of water, which inhibits further reaction progress and reduces overall conversion efficiency.

The conceptual design of the reactive distillation column integrates a reactive zone for the esterification reactions with separation zones for selective removal of water and intermediate esters. This design ensures continuous removal of inhibitory by-products, enhancing both conversion rates and product purity.

Case study description

The case study evaluates a feed stream containing 90% citric acid and 10% ethanol by weight, with a total feed flowrate of 1000 kg/h. The feed enters the reactive distillation column at the middle stage, designed to optimize the distribution of reactants and products across the reactive and separation zones. The column operates under vacuum conditions at 0.1 atm, which reduces the boiling point of water and facilitates its removal, enhancing the overall reaction equilibrium. The column contains 50 stages, with 25 allocated to the reactive zone and the remainder for the separation zones. This configuration was selected to maximize resource utilization and product recovery. In the same sense, a reactive DWC was considered in order to have an alternative. Fig. 6.14 shows both alternatives considered for this case study (Santaella et al., 2017).

The optimization of the reactive DWC was performed using a Differential Evolution algorithm combined with Tabu Search (DETS). The objective of the optimization was to minimize the TAC while ensuring compliance with product purity (>99%) and maximizing the conversion efficiency of citric acid to TEC. The optimization considered several variables, including the reflux ratio, feed stage location, tray configuration, and reboiler duty. The function objective balanced economic considerations with operational parameters such as energy efficiency and conversion rates. This multi-objective approach allowed for a comprehensive evaluation of design and operational variables to identify the most effective configuration (Srinivas & Rangaiah, 2007).

Results

The reactive DWC achieved a 35% reduction in energy consumption compared to conventional systems, driven by its ability to combine reaction and separation within a single unit. The TAC was reduced by 20%, reflecting savings in both operational and equipment costs. Design trends indicated that higher tray counts in the reactive zone improved conversion rates, while optimal ethanol-to-citric acid ratios minimized unreacted intermediates. Additionally, the removal of water as a by-product was more effective in the

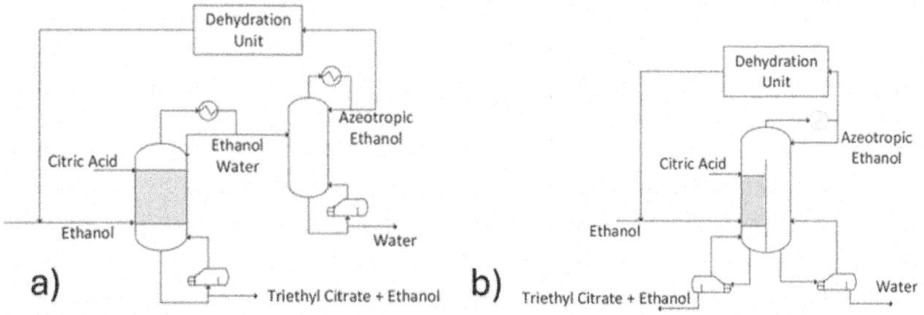

FIGURE 6.14 RC (a) and RDWC (b) schemes for TEC production. *Dashed area* indicates reactive section.

TABLE 6.4 Design parameters for both optimal alternatives, DWC and reactive DWC.

Variable	RD	RDWC
EtOH stream (kg/h)	1771	1637
CA stream (kg/h)	1801	1812
Heating duty column 1 (prefractionator) (kW)	717	710
Heating duty column 2 (kW)	998	366
Reflux ratio column 1 (prefractionator)	0.01	0.8
Reflux ratio column 2	0.86	–
TEC (kg/h)	1291	1291
Number of stages column 1 (prefractionator)	60	70
Number of stages column 2	20	38
Number of reactive stages	57	56
Conversion of CA (%)	100	100
Selectivity to TEC (%)	99.7	99.1
Energy consumption (kJ/kg TEC)	4780	3003

DWC, reducing reaction inhibition and increasing overall efficiency. The results from the optimization study highlighted the superior performance of the reactive DWC compared to conventional and thermally coupled configurations. Table 6.4 shows the design parameters for both optimal alternatives, DWC and reactive DWC.

Case study: Production of FAMEs

Fatty acid methyl esters (FAMEs) are esters derived from fatty acids and methanol, widely recognized as the primary constituents of biodiesel. FAMEs play a critical role in the global push toward sustainable energy solutions as they serve as a renewable and biodegradable alternative to conventional fossil fuels. Their use reduces greenhouse gas emissions, promotes energy independence, and supports the circular economy by utilizing waste oils and fats as feedstocks. The production of FAMEs has become a cornerstone of biofuel development, particularly in addressing the energy demands of transportation and industry (Lores et al., 2006). The production of FAMEs involves the esterification of free fatty acids (FFAs) and the transesterification of triglycerides with methanol, catalyzed by acidic or basic catalysts. Traditional production methods often rely on batch reactors and involve multiple steps, including separation and purification, which are energy-intensive and generate substantial waste. From a sustainability perspective, integrating reaction and separation into a single unit, such as a reactive distillation column or a reactive DWC, offers a pathway to minimize energy consumption, improve reaction efficiency, and reduce waste. The objective of this section is to evaluate an optimized reaction-separation alternative for FAME production using thermally coupled systems and reactive DWCs, with a particular focus on assessing the TAC.

Esterification and conceptual design

The production of FAMEs through esterification involves the reaction of FFAs with methanol, a process governed by equilibrium and kinetics (Vyas et al., 2010):

$$RCOOH + CH_3OH \leftrightarrow RCOOCH_3 + H_2O$$

This reaction is equilibrium-limited, with water formation inhibiting the forward reaction. The RDWC provides a solution by simultaneously facilitating the reaction and removing water as a by-product through a side stream. Methanol is recovered as a distillate and recycled, while FAMEs are obtained as a high-purity bottom product. The design of the RDWC was performed using Aspen Plus, employing the UNIFAC-Dortmund thermodynamic model for accurate property estimation. The feed stream consisted of 1000 kg/h of a mixture containing 90% FFAs and 10% methanol, with a molar ratio of 1:6. The RDWC included 40 total stages, with 15 reactive stages strategically placed to optimize reactant conversion and by-product removal (Kiss, 2014).

Optimization procedure

To identify the optimal design, a simulated annealing (SA) algorithm was implemented in MATLAB, coupled with rigorous Aspen Plus simulations (Delahaye et al., 2018). The optimization focused on minimizing the reboiler duty as a proxy for TAC, considering variables such as Number and location of reactive stages, Reflux ratio, Liquid and vapor split ratios, Side stream flow rate.

The objective function incorporated energy consumption and purity constraints, ensuring compliance with product specifications (>99.9% FAME purity). This approach enabled a comprehensive evaluation of design and operational parameters to achieve global optimality.

Results and discussion

The RDWC demonstrated superior performance compared to conventional biodiesel production processes. The key findings are summarized below:

- Energy efficiency: The RDWC achieved a 25% reduction in energy consumption compared to the CMP, with a total heating duty of 382.62 kW.

- Cost savings: The TAC for the RDWC was $179,120/year, significantly lower than the base case, reflecting reduced equipment and operational costs.
- Environmental impact: The optimized RDWC design reduced greenhouse gas emissions by 25%, with annual CO_2 emissions of 1012 tons, compared to 1405 tons in the base case.

Increasing the number of reactive stages significantly enhanced conversion efficiency and improved water removal, which played a crucial role in achieving higher FAME purity levels. Additionally, the optimization of liquid and vapor split ratios minimized remixing within the column, thereby improving energy utilization and overall process performance. These advancements in design and operation also translated into notable environmental benefits, as the reduced energy requirements directly contributed to lower CO_2 emissions, reinforcing the RDWC's sustainability advantages in FAME production. Table 6.5 shows the main parameters design for the RDWC showed in Fig. 6.15.

Case study: Pilot plant scale dividing wall column

Distillation remains a cornerstone of industrial separation processes due to its versatility and capability to handle large-scale operations. However, the energy-intensive nature of traditional distillation systems has driven the development of advanced technologies, including DWCs and reactive dividing wall columns (RDWCs) (Kiss et al., 2009). These innovative designs integrate separation and, when applicable, reaction processes into a single unit, offering substantial reductions in energy consumption and operating costs. As explored in previous sections, DWCs have demonstrated superior performance for complex separations, while RDWCs further extend these benefits to processes involving simultaneous reaction and separation. The objective of this section is to provide detailed insights into the design, operation, and performance of a pilot-scale RDWC implemented at the Department of Chemical Engineering at the University of Guanajuato, serving as proof-of-concept for industrial applications.

System analyzed

The pilot-scale RDWC was designed to separate a ternary mixture of methanol, ethanol, and water—an azeotropic system characterized by closely boiling components, posing significant challenges to conventional separation methods (Barroso-Muñoz et al., 2011). The system was engineered with a total height of 6.5 m and an internal diameter of 0.1 m, operating at atmospheric pressure. The column was equipped with 30 stages, divided into three functional sections: the rectifying section, the prefractionator section, and the stripping section. A unique feature of this design is the dividing wall, which

TABLE 6.5 Parameter design for the RDWC.

Parameter	RDWC
Number of stages, N	36
Number of reactive stages, NR	21
Liquid split stage, NL	29
Vapor split stage, NV	13
Side-draw stage, N1	6
Organic phase return stage, N2	26
Acid flowrate (kmol/h)	5.83
Temperature HEX1 (K)	418.9
Methanol flowrate (kmol/h)	6.997
Temperature HEX2 (K)	338.7
Distillate rate (kg/h)	37.12
Boilup ratio (kg/kg)	3.39
Side draw flowrate (kg/h)	122.04
Liquid flowrate, L1 (kmol/h)	1.29
Vapor flowrate, V1 (kmol/h)	2.14
Water flowrate (kmol/h)	5.67
Water purity (%)	96.99
FAME flowrate (kmol/h)	5.83
FAME purity (wt%)	99.96
Recycle stream TOPALC flowrate (kmol/h)	1.07
TOPALC purity (%)	92.39
Energy requirements HEX1 duty (kW)	95.88
Energy requirements HEX2 duty (kW)	9.54
Reboiler duty (kW)	200.72
Condenser duty (kW)	−155.18
Net heating duty (kW)	355.89
Total cooling duty (kW)	−355.89
Key performance indicators (kWh/ton FAME)	306.1
CO_2 emission (tons/year)	1012

Continued

TABLE 6.5 Parameter design for the RDWC.—cont'd

Parameter	RDWC
Annual operating cost (k$/year)	107.71
Capital cost (k$)	25.06
Total annual cost (k$/year)	179.12

splits the prefractionator section into two distinct zones. This configuration enables efficient separation by minimizing remixing and reducing energy losses. The feed, consisting of 50 mol% methanol, 30 mol% ethanol, and 20 mol% water, is introduced at stage 15 at a flow rate of 10 kg/h. Methanol is recovered at the distillate with high purity, ethanol is drawn as a side product, and water is extracted at the bottom of the column. Instrumentation for the RDWC included temperature sensors and flow meters to monitor operational stability and ensure precise control of the separation process (Barroso-Muñoz et al., 2011). The system was supported by a total condenser and reboiler, completing the required energy integration for efficient operation (see Fig. 6.16).

The energy-efficient design of the RDWC

The RDWC was designed with a focus on energy efficiency and optimal separation performance. The design process was conducted using Aspen Plus, leveraging the UNIQUAC thermodynamic model to accurately predict phase behavior and azeotropic interactions in the methanol-ethanol-water mixture.

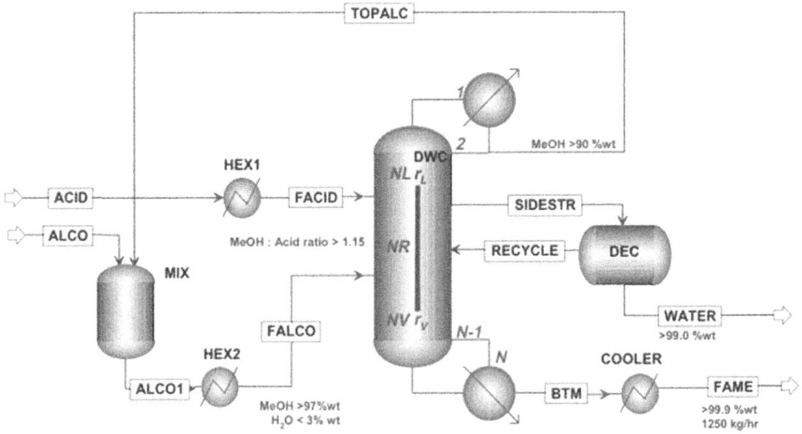

FIGURE 6.15 Schematic process for FAME production using an RDWC.

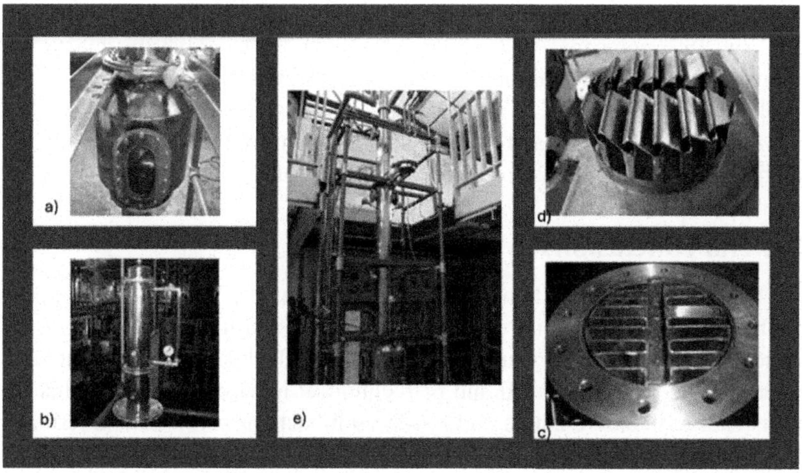

FIGURE 6.16 Physical implementation of an RDWC; (a) reboiler, (b) condenser, (c) distributor tray, (d) support tray, (e) final implementation.

This model provided the foundation for rigorous simulation and optimization of the column configuration. The design optimization involved determining the optimal number of stages in each section, the feed stage location, and the liquid and vapor split ratios at the dividing wall. Specific attention was given to the reactive and separation zones to enhance the removal of water and unreacted components, thereby driving the equilibrium toward the desired products. The feed composition and flow rate were carefully calibrated to reflect real-world operational scenarios, ensuring the pilot plant's relevance to industrial applications (Hernández & Jiménez, 1999).

Energy efficiency was a key driver of the design process. The RDWC minimized reboiler duty by integrating the dividing wall, which allowed for heat and mass transfer efficiencies not achievable in conventional systems. This resulted in substantial energy savings, with a measured reboiler duty of 3.8 kW, representing a 30% reduction compared to traditional two-column configurations.

Results

The pilot-scale RDWC exhibited superior performance, achieving significant energy savings and high product purity. The system reduced energy consumption by 30% compared to a conventional two-column setup, with a measured reboiler duty of 3.8 kW. Methanol purity in the distillate stream reached 99.5 mol%, ethanol purity in the side product stream was 97 mol%, and water purity in the bottoms stream exceeded 99 mol%. The system operated stably across a range of feed compositions and flow rates, demonstrating its robustness and adaptability to dynamic conditions. Economic analysis revealed notable cost benefits, with the integrated design lowering the TAC by 28% compared to traditional configurations, primarily due to the elimination of intermediate equipment like reboilers and condensers. These results validate the RDWC's potential for scaling up to industrial applications, providing a sustainable and economically advantageous solution for challenging separation processes (Barroso-Muñoz et al., 2011).

General conclusions

The advancements in distillation technology presented in this chapter highlight the transformative potential of DWCs) and RDWCs in addressing the challenges of energy efficiency and process sustainability. Traditional distillation, a cornerstone of the chemical and petrochemical industries, is often hindered by its energy-intensive nature and high operational costs. DWCs and RDWCs offer an innovative alternative by integrating separation and reaction processes into a single unit, delivering substantial energy savings, cost reductions, and environmental benefits. Their ability to handle complex separations, including azeotropic and multicomponent systems, was demonstrated through detailed case studies, showcasing operational flexibility and superior efficiency where conventional configurations falter. This chapter also explored the unique advantages of RDWCs in reactive systems, exemplified by the production of high value-added compounds such as triethyl citrate and fatty acid methyl esters (FAMEs). The integration of reaction and separation in RDWCs improves conversion efficiency, minimizes equipment requirements, and simplifies operational complexity, paving the way for more sustainable industrial practices. Additionally, the pilot-scale RDWC implemented at the University of Guanajuato provided valuable insights into the design, optimization, and operation of advanced distillation systems. The pilot plant demonstrated the scalability and practicality of RDWCs, achieving a 30% reduction in energy consumption alongside significant economic savings when compared to traditional configurations. The study further emphasized the importance of rigorous design methodologies, facilitated by Aspen Plus simulations and the application of robust thermodynamic models like UNIQUAC, to ensure accurate system predictions and reliable performance.

Throughout this chapter, several key themes emerged. DWCs and RDWCs consistently demonstrated their capacity to reduce energy consumption by lowering reboiler duties while maintaining or exceeding separation performance. Their economic advantages, including reduced equipment costs and lower TAC, position these technologies as financially viable solutions across diverse industrial applications. The operational stability of DWCs and RDWCs, even under varying feed compositions and operating conditions, underscores their versatility and adaptability for complex separation challenges. Furthermore, their ability to reduce energy demands and associated CO_2 emissions significantly enhances their sustainability credentials, aligning with global goals for greener chemical processes. The findings and case studies discussed in this chapter illustrate how DWCs and RDWCs can revolutionize separation processes in the chemical industry. By bridging the gap between efficiency and sustainability, these technologies represent a paradigm shift toward intensified and environmentally conscious process design. The continued development and implementation of DWCs and RDWCs will be essential in meeting the growing demand for energy-efficient and cost-effective industrial solutions, setting the stage for a more sustainable future in chemical engineering.

References

Asprion, N., & Kaibel, G. (2010). Dividing wall columns: Fundamentals and recent advances. *Chemical Engineering and Processing: Process Intensification, 49*(2), 139–146.

Barroso-Muñoz, F. O., Hernández, S., Segovia-Hernández, J. G., Hernández-Escoto, H., Rico-Ramírez, V., & Chávez, R.-H. (2011). Implementation and operation of a dividing-wall distillation column. *Chemical Engineering and Technology, 34*(5), 746–750. https://doi.org/10.1002/ceat.201000388

Bravo, J. L. (2019). *Distillation technology: What's next?* AIChE.

Canton, H. (2021). *International energy agency—IEA*. Informa UK Limited. https://doi.org/10.4324/9781003179900-103

Contreras-Zarazúa, G., Martin-Martin, M., Sánchez-Ramirez, E., & Segovia-Hernández, J. G. (2022). Furfural production from agricultural residues using different intensified separation and pretreatment alternatives. Economic and environmental assessment. *Chemical Engineering and Processing - Process Intensification, 171*, Article 108569. https://doi.org/10.1016/j.cep.2021.108569

Contreras-Zarazúa, G., Sánchez-Ramírez, E., Vázquez-Castillo, J. A., Ponce-Ortega, J. M., Errico, M., Kiss, A. A., & Segovia-Hernández, J. G. (2019). Inherently safer design and optimization of intensified separation processes for furfural production. *Industrial and Engineering Chemistry Research, 58*(15), 6105–6120. https://doi.org/10.1021/acs.iecr.8b03646

Dünnebier, G., & Pantelides, C. C. (1999). Optimal design of thermally coupled distillation columns. *Industrial and Engineering Chemistry Research, 38*(1), 162–176. https://doi.org/10.1021/ie9802919

Dejanović, I., Jansen, H., & Olujić, Ž. (2014). Dividing wall column as energy saving retrofit technology. In *10th International conference on distillation & absorption* (pp. 625–630).

Dejanović, I., Matijašević, L., Jansen, H., & Olujić, Ž. (2011). Designing a packed dividing wall column for an aromatics processing plant. *Industrial and Engineering Chemistry Research, 50*(9), 5680–5692. https://doi.org/10.1021/ie1020206

Dejanović, I., Matijašević, L. J., & Olujić, Ž. (2010). Dividing wall column—A breakthrough towards sustainable distilling. *Chemical Engineering and Processing: Process Intensification, 49*(6), 559–580. https://doi.org/10.1016/j.cep.2010.04.001

Delahaye, D., Chaimatanan, S., & Mongeau, M. (2018). Simulated annealing: From basics to applications. In *Handbook of metaheuristics* (pp. 1–35). Cham: Springer International Publishing.

Delolo, F. G., Dos Santos, E. N., & Gusevskaya, E. V. (2019). Anisole: A further step to sustainable hydroformylation. *Green Chemistry, 21*(5), 1091–1098. https://doi.org/10.1039/c8gc03750g

Errico, M., Sanchez-Ramirez, E., Quiroz-Ramìrez, J. J., Rong, B. G., & Segovia-Hernandez, J. G. (2017). Multiobjective optimal acetone-butanol-ethanol separation systems using liquid-liquid extraction-assisted divided wall columns. *Industrial and Engineering Chemistry Research, 56* (40), 11575–11583. https://doi.org/10.1021/acs.iecr.7b03078

Fang, J., Zhao, H., Qi, J., Li, C., Qi, J., & Guo, J. (2015). Energy conserving effects of dividing wall column. *Chinese Journal of Chemical Engineering, 23*(6), 934–940. https://doi.org/10.1016/j.cjche.2014.08.009

Finkelstein, M., & Gold, H. (1959). Toxicology of the citric acid esters: tributyl citrate, acetyl tributyl citrate, triethyl citrate, and acetyl triethyl citrate. *Toxicology and Applied Pharmacology, 1*(3), 283–298. https://doi.org/10.1016/0041-008x(59)90113-9

Górak, A., & Olujić, Ž. (2014). *Distillation: Equipment and processes*. Germany: Elsevier Inc.. https://doi.org/10.1016/C2010-0-67738-8

Hernández, S., & Jiménez, A. (1999). Design of energy-efficient Petlyuk systems. *Computers and Chemical Engineering, 23*(8), 1005–1010. https://doi.org/10.1016/s0098-1354(99)00257-4

Hoyme, C. A. (2017). *Dividing wall columns in the chemical industry.* Springer Science and Business Media LLC. https://doi.org/10.1007/978-3-319-52287-6_38

Ignat, R. M., & Kiss, A. A. (2013). Optimal design, dynamics and control of a reactive DWC for biodiesel production. *Chemical Engineering Research and Design, 91*(9), 1760–1767. https://doi.org/10.1016/j.cherd.2013.02.009

Kalita, R., & Gentry, J. C. (2018). Dividing wall columns in operation. *Chemical Engineering Transactions, 69*, 547–552. https://doi.org/10.3303/CET1869092

Kiss, A. A. (2014). Distillation technology–still young and full of breakthrough opportunities. *Journal of Chemical Technology & Biotechnology, 89*(4), 479–498.

Kiss, A. A., & Ignat, R. M. (2012). Enhanced methanol recovery and glycerol separation in biodiesel production - DWC makes it happen. *Applied Energy, 99*, 146–153. https://doi.org/10.1016/j.apenergy.2012.04.019

Kiss, A. A., Pragt, J. J., & van Strien, C. J. G. (2009). Reactive dividing-wall columns-how to get more with less resources? *Chemical Engineering Communications, 196*(11), 1366–1374. https://doi.org/10.1080/00986440902935507

Kolah, A. K., Asthana, N. S., Vu, D. T., Lira, C. T., & Miller, D. J. (2008). Triethyl citrate synthesis by reactive distillation. *Industrial and Engineering Chemistry Research, 47*(4), 1017–1025. https://doi.org/10.1021/ie070279t

Lorenz, H. M., Staak, D., Grützner, T., & Repke, J. U. (2018). Divided wall columns: Usefulness and challenges. *Chemical Engineering Transactions, 69*, 229–234. https://doi.org/10.3303/CET1869039

Lores, M., Gómez-Brandón, M., Pérez-Díaz, D., & Domínguez, J. (2006). Using FAME profiles for the characterization of animal wastes and vermicomposts. *Soil Biology and Biochemistry, 38*(9), 2993–2996. https://doi.org/10.1016/j.soilbio.2006.05.001

Petlyuk, F. (1965). *Thermodynamically optimal method for separating multicomponent mixtures.*

Piszczek, R., Hergenrother, M., & Nahigian, G. (2023). Distillation improvement opportunities part 2: Improvements through the use of dividing wall columns. *TCE: The Chemical Engineer*, (984).

Rév, E., Emtir, M., Szitkai, Z., Mizsey, P., & Fonyó, Z. (2001). Energy savings of integrated and coupled distillation systems. *Computers and Chemical Engineering, 25*(1), 119–140. https://doi.org/10.1016/s0098-1354(00)00643-8

Rong, B.-G. (2011). Synthesis of dividing-wall columns (DWC) for multicomponent distillations—A systematic approach. *Chemical Engineering Research and Design, 89*(8), 1281–1294. https://doi.org/10.1016/j.cherd.2011.03.014

Suphanit, B., Bischert, A., & Narataruksa, P. (2007). Exergy loss analysis of heat transfer across the wall of the dividing-wall distillation column. *Energy, 32*(11), 2121–2134. https://doi.org/10.1016/j.energy.2007.04.006

Santaella, M. A., Jiménez, L. E., Orjuela, A., & Segovia-Hernández, J. G. (2017). Design of thermally coupled reactive distillation schemes for triethyl citrate production using economic and controllability criteria. *Chemical Engineering Journal, 328*, 368–381. https://doi.org/10.1016/j.cej.2017.07.015

Schultz, M. A., Stewart, D. G., Harris, J. M., Rosenblum, S. P., Shakur, M. S., & Brien, D. (2002). *Reduce costs with dividing wall columns.*

Seider, W. D., Seader, J. D., & Lewin, D. R. (1998). *Process design principles: Synthesis, analysis and evaluation chemistry.*

Sholl, D. S., & Lively, R. P. (2016). Seven chemical separations to change the world. *Nature, 532*(7600), 435–437. https://doi.org/10.1038/532435a

Srinivas, M., & Rangaiah, G. P. (2007). Differential evolution with tabu list for solving nonlinear and mixed-integer nonlinear programming problems. *Industrial and Engineering Chemistry Research, 46*(22), 7126–7135. https://doi.org/10.1021/ie070007q

Sundmacher, K., & Kienle, A. (2003). *Reactive distillation: Status and future directions.* Wiley. https://doi.org/10.1002/3527600523

Tinoco-Sáenz, R., Alcantara-Avila, J. R., Mansouri, S. S., Sánchez-Ramírez, E., Ramírez-Márquez, C., & Segovia-Hernández, J. G. (2022). Design and optimization of an inherently safe and sustainable process for the separation of anisole. *Journal of Loss Prevention in the Process Industries, 80.* https://doi.org/10.1016/j.jlp.2022.104899

Triantafyllou, C., & Smith, R. (1992). Design and optimisation of fully thermally coupled distillation columns. *Chemical Engineering Research and Design, 70*(2), 118–132.

Triantafyllou, C., & Smith, R. (1993). *The design and optimization of dividing wall distillation columns.* Springer Science and Business Media LLC. https://doi.org/10.1007/978-94-011-1454-7_32

Vyas, A. P., Verma, J. L., & Subrahmanyam, N. (2010). A review on FAME production processes. *Fuel, 89*(1), 1–9. https://doi.org/10.1016/j.fuel.2009.08.014

Yildirim, O., Kiss, A. A., & Kenig, E. Y. (2011). Dividing wall columns in chemical process industry: A review on current activities. *Separation and Purification Technology, 80*(3), 403–417. https://doi.org/10.1016/j.seppur.2011.05.009

Zhu, X., Mallinson, R. G., & Resasco, D. E. (2010). Role of transalkylation reactions in the conversion of anisole over HZSM-5. *Applied Catalysis A: General, 379*(1–2), 172–181. https://doi.org/10.1016/j.apcata.2010.03.018

Zuber, S. A. N. A., Rusli, A., & Ismail, H. (2019). Effectiveness of triacetin and triethyl citrate as plasticizer in polyvinyl alcohol. In *Materials today: Proceedings* (vol. 17, pp. 560–567). Elsevier Ltd. https://doi.org/10.1016/j.matpr.2019.06.335. https://www.sciencedirect.com/journal/materials-today-proceedings.

Chapter 7

Methodologies for the synthesis and design of DWC type schemes

Motivation and theoretical aspects of DWCs with more than one wall

A Dividing Wall Column (DWC) is a highly efficient configuration for distillation that integrates the functionality of two columns within a single shell, separated by a vertical dividing wall. Thermodynamically equivalent to the Petlyuk column, the DWC offers substantial energy savings by eliminating the need for additional reboilers and condensers. It achieves this by performing two thermally coupled separations simultaneously: the lightest component is removed as the distillate, the heaviest component is collected as the bottom product, and an intermediate component is extracted via a side draw. This configuration significantly enhances operational efficiency and reduces capital and energy costs (Halvorsen & Skogestad, 2003).

Mathematical model of the DWC

The mathematical modeling of a DWC involves extending the conventional MESH equations (Material balance, Equilibrium relations, Summation equations, and Enthalpy balance) to include the effects of pressure drop. This augmented framework is often referred to as the MESHD model. Below, the key equations of the MESHD model and the physical meaning of the variables are presented (Sangal et al., 2012).

- Material balance equations

  For any stage i and component j in the DWC:

$$l_{i,j} + v_{i-1,j} = v_{i,j} + l_{i+1,j}$$

where:

$l_{i,j}$: Molar flow rate of component j in the liquid phase leaving stage i (kmol/h)

Thermally Coupled Distillation Columns. https://doi.org/10.1016/B978-0-443-33114-5.00006-5

$v_{i-1,j}$: Molar flow rate of component j in the vapor phase leaving stage i (kmol/h).

At the condenser ($i = 1$) and reboiler ($i = N$):

$$\sum_{j=1}^{c} l_{N,j} = B, \sum_{j=1}^{c} v_{1,j} = D$$

where:

B: Bottom product flow rate (kmol/h)
D: Distillate flow rate (kmol/h)
For the side draw (S_3) at stage $i = N_s3$:

$$S_3 = \sum_{j=1}^{c} l_{i,j}$$

- Equilibrium relations

The vapor-liquid equilibrium (VLE) is expressed as:

$$y_{i,j} = K_{i,j} \quad x_{i,j}$$

where:

$y_{i,j}$: Mole fraction of component j in the vapor phase
$x_{i,j}$: Mole fraction of component j in the liquid phase
$K_{i,j}$: Equilibrium constant for component j, which is a function of temperature (T_i) and pressure (P_i).

The equilibrium constant is defined by thermodynamic models, such as Peng-Robinson or NRTL, ensuring consistency between the phases.

- Enthalpy balance equations

For each stage i, the energy balance is:

$$L_i h_i + V_{i-1} \quad H_{i-1} = L_{i+1} h_{i+1} + V_i H_i + Q_i$$

where:

L_i: Total liquid flow leaving stage i (kmol/h)
V_i: Total vapor flow leaving stage i (kmol/h)
h_i, H_i: Specific enthalpy of the liquid and vapor streams, respectively, (kJ/kmol)
Q_i: Heat duty at stage i (kJ/h).

The enthalpy terms are derived from integrating heat capacities and include corrections for phase changes and departure from ideal gas behavior.

- Pressure balance equations

The pressure drop (ΔP) between consecutive trays is:

$$P_{i-1}P_i = \Delta P_i = f\ (L_i V_i, \text{tray} \quad \text{design} \quad \text{parameters})$$

where:

P_i: Pressure at stage i (kPa)
f: Function of tray geometry (e.g., weir height, and hole diameter) and flow rates.

The DWC model assumes equal pressure drops across the two sections of the divided wall to maintain hydraulic balance. This constraint affects the vapor and liquid splits at the interface of the divided sections.

- Split ratios

In the divided section:

$$R_1 = \frac{L_{S1}}{(L_{S1} + L_{S2})}, \quad R_2 = \frac{V_{S2}}{(V_{S1} + V_{S2})}$$

where:

R_1: Liquid split ratio,
R_2: Vapor split ratio
L_{S1}, L_{S2}: Liquid flow rates entering sections 1 and 2, respectively
V_{S1}, V_{S2}: Vapor flow rates entering sections 1 and 2, respectively.

These ratios determine the distribution of mass and energy between the two sections and are crucial for achieving the desired product specifications.

- Physical implications and model assumptions

Thermal coupling: The DWC assumes perfect thermal coupling between the divided sections. This implies equal pressure drops and balanced energy distribution.

Hydraulic constraints: The model accounts for the unique hydraulic behavior of the DWC, where the liquid and vapor splits are naturally adjusted to maintain pressure balance across the divided wall.

Steady-state operation: The equations are solved under steady-state conditions, ignoring transient dynamics.

Ideal Stage assumption: Each tray is assumed to achieve equilibrium, simplifying the calculation of phase interactions.

Negligible axial mixing: The model assumes negligible mixing along the axial direction within each section.

- Degrees of freedom analysis

For a DWC with C components, N_T trays, and three product streams: (8)

$$\text{Degrees of Freedom} = 2\text{N\_T}\ (\text{C} + 1) + \text{C} + 6 - \text{Independient} \qquad (7.1)$$

After fixing feed conditions ($C + 2$) and specifying column pressure (N_T), four degrees of freedom remain. These typically include Reflux ratio (R), Liquid split ratio (R_1), Side draw flow rate (S_3), and Reboiler duty (Q_B).

This balance ensures the model's solvability and convergence during simulation. The DWC model integrates a detailed representation of mass, energy, and momentum balances, augmented with split ratios and pressure drop equations. By leveraging these equations, engineers can optimize the design and operation of DWCs to achieve energy-efficient separations.

Energy implication of the interconnection flows in the thermal couplings of dividing wall columns

The DWC is a revolutionary technology in distillation processes that integrates thermal coupling to optimize energy use and separation efficiency. At the heart of this innovation lies the concept of interconnection flows—the streams of vapor and liquid redistributed between the prefractionator and main column sections. These flows are fundamental to the DWC's energy-conserving capabilities, ensuring the minimization of reboiler and condenser duties while simultaneously improving separation quality and product purities. This section delves into the multifaceted implications of interconnection flows, encompassing their role in reducing energy requirements, enhancing thermodynamic and exergetic performance, and improving the operational and separation efficiency of the column.

Energy savings and thermal coupling

The redistribution of interconnection flows in DWCs eliminates the redundancy of auxiliary reboilers and condensers typically required in conventional distillation column setups. By channeling energy directly from one section of the column to another, interconnection flows enable the DWC to operate as a thermally coupled system, significantly reducing the reboiler and condenser duties. The optimized vapor and liquid split ratios (R_2 and R_1) allow the energy provided by the reboiler to be utilized more effectively, ensuring that the energy is consumed only where it is required for component separation. Studies report energy savings ranging from 30% to 50% compared to traditional distillation systems, making DWCs an attractive solution in industries with high energy costs (Halvorsen & Skogestad, 2003).

A key aspect of this energy reduction is the role of heat transfer across the dividing wall, particularly in heat transfer dividing wall columns (HTDWCs). The temperature gradient across the dividing wall creates opportunities for positive heat transfer, where energy from one side of the wall is reused on the other, enhancing thermodynamic efficiency. However, the direction and magnitude of heat transfer must be carefully managed, as adverse heat transfer can lead to energy inefficiencies. The design and operational parameters of interconnection flows are thus critical to harnessing the full energy-saving potential of the DWC.

Thermodynamic and exergetic performance

From a thermodynamic perspective, DWCs demonstrate superior efficiency due to their ability to minimize entropy generation. The interconnection flows reduce the irreversible losses associated with heat and mass transfer, contributing to a lower overall energy requirement for the separation process. The thermodynamic efficiency (η) of a DWC is often expressed as the ratio of the minimum work required for separation to the actual energy input. By redistributing energy within the column, interconnection flows enable the DWC to approach this theoretical minimum more closely than conventional systems (Suphanit et al., 2007).

Exergetic analysis further underscores the benefits of interconnection flows. Exergy, which represents the useful portion of energy in a thermodynamic process, is conserved more effectively in DWCs due to the elimination of back-mixing and the optimized use of energy. This is particularly evident in processes involving close-boiling mixtures, where traditional systems suffer from significant energy wastage. By directing vapor and liquid streams to where they are most needed, interconnection flows ensure that the exergy destruction is minimized, thereby enhancing the overall process sustainability (Ignat & Kiss, 2013).

Separation efficiency and product quality

Beyond energy and thermodynamic considerations, interconnection flows play a pivotal role in enhancing the separation performance of DWCs. By distributing the components of a mixture more effectively between the prefractionator and the main column, these flows allow for sharper separations and higher product purities. The prefractionator section performs an initial split between light and heavy components, while intermediate components are directed to the main column for further separation. This multi-effect separation minimizes remixing, a common issue in conventional systems that reduces separation sharpness and increases energy consumption (Fang et al., 2015).

The role of interconnection flows is particularly pronounced in processes involving ternary mixtures. For example, in the separation of mixtures containing light (A), intermediate (B), and heavy (C) components, the prefractionator initially separates A and C while allowing B to distribute between streams. The main column then completes the separation, yielding highly pure products in the distillate, bottom, and side-draw streams. This efficient separation mechanism is a direct result of the interconnection flows that ensure proper energy and material distribution throughout the column.

Interconnection flows also enable DWCs to handle azeotropic and close-boiling mixtures more effectively. By introducing a controlled energy gradient across the dividing wall, these flows facilitate the disruption of azeotropes, allowing for their separation without the need for additional entrainers or complex column configurations. This capability extends the applicability of DWCs to a wider range of industrial processes, including those involving highly nonideal mixtures (Flores et al., 2003).

Implications for modeling and design

The incorporation of interconnection flows into the modeling and design of DWCs introduces significant complexity but is essential for accurate representation and optimization of the column's performance. The mathematical representation of DWCs is based on MESHD equations, which account for material balances, equilibrium relations, summation equations, enthalpy balances, and pressure drops. The inclusion of interconnection flows adds new variables, such as split ratios (R_1 and R_2) and heat transfer coefficients, which must be carefully defined to capture the interactions between different sections of the column (Ho et al., 2011).

Dynamic modeling of DWCs further complicates the analysis, as interconnection flows exhibit transient behavior during startup and operational changes. Ensuring stability and efficiency under such conditions requires precise control strategies, often involving advanced process control techniques such as model predictive control. These strategies enable real-time adjustment of interconnection flows to accommodate changes in feed composition, product specifications, or operating conditions (Buck et al., 2011).

From a design perspective, the placement and sizing of the dividing wall are critical to the effective distribution of interconnection flows. Factors such as tray configuration, weir height, and packing materials must be optimized to maintain hydraulic and thermal balance within the column. Structured packing is often preferred for its low-pressure drop characteristics, which facilitate smoother flow distribution and improved separation efficiency. Additionally, the thermal conductivity of the dividing wall material must be carefully selected to enable positive heat transfer while minimizing adverse effects (Lestak & Collins, 1997).

Operational implications and flexibility

Operationally, interconnection flows provide DWCs with unparalleled flexibility, allowing for adjustments to vapor and liquid splits to suit varying feed compositions and separation requirements. This adaptability is particularly valuable in industries with dynamic production needs, where feedstocks or product demands may change frequently. The ability to fine-tune interconnection flows ensures that the DWC can maintain optimal performance across a wide range of operating conditions.

Moreover, interconnection flows offer opportunities for process intensification, enabling the integration of additional functionalities such as reaction or heat integration within the same column. This multi-functionality reduces equipment costs and footprint, further enhancing the economic and environmental benefits of DWCs (Christiansen et al., 1997).

Thus, Interconnection flows are the cornerstone of the energy efficiency and operational excellence of DWC. By redistributing energy and material within the column, these flows reduce energy consumption, enhance thermodynamic and exergetic performance, and improve separation quality. Their implications extend to the modeling, design, and operation of DWCs, underscoring their significance in achieving optimal column performance. As industries continue to prioritize sustainability and efficiency, the role of interconnection flows in DWCs will remain a critical area of research and innovation.

Design strategies for dividing wall columns

The design of DWCs is a cornerstone of modern distillation technology, known for its potential to achieve substantial energy savings and superior separation performance. Unlike conventional distillation columns, DWCs integrate the prefractionator and the main column within a single shell, separated by a vertical dividing wall. This innovative configuration not only reduces the need for multiple reboilers and condensers but also enables sharper separations, making DWCs highly efficient and cost-effective. However, the inherent complexity of DWCs requires robust design strategies to optimize their performance and fully exploit their benefits. Key challenges in the design of DWCs include the need to balance interconnection flows, manage hydraulic interactions, and optimize thermal coupling across the dividing wall. These factors are critical for maintaining the column's operational stability and achieving the desired separation targets. Advanced simulation tools, combined with thermodynamic and heuristic methods, have emerged as indispensable resources for addressing these challenges. By incorporating detailed thermodynamic models and pressure drop correlations, designers can better predict the behavior of DWCs under varying operating conditions, enabling more accurate and efficient designs.

This section aims to present a comprehensive overview of methodologies available for the design of DWCs. The discussion will highlight the importance of adopting appropriate design strategies to address the unique challenges posed by DWCs and ensure their optimal operation. Through the exploration of these methods, this section seeks to provide a practical guide for the conceptualization and implementation of high-performing DWCs, supported by proven approaches and scientific insights (Buitimea-Cerón et al., 2021).

Triantafyllou & Smith methodology

The Triantafyllou & Smith (T&S) methodology is a foundational approach for the design of DWCs, extensively used due to its practicality and adaptability. Developed based on the principles of the Fenske-Underwood-Gilliland-Kirkbride (FUGK) equations, this methodology conceptualizes a DWC by initially representing it as a simplified three-column system analogous to the Petlyuk arrangement. The prefractionator and main column sections are modeled separately, enabling a systematic estimation of key operational parameters, including the number of stages, reflux ratio, and interconnection flows. This approach offers an effective balance between simplified conceptual design and the rigorous computations required for real-world implementation.

The methodology begins with the shortcut modeling of the three-column system. This is typically performed using tools such as the DSTWU (Distillation Tower Unit) in Aspen Plus or similar software modules. The prefractionator is the focus of initial calculations, with its reflux ratio set at 1.2 times the minimum value to ensure a conservative starting point. The DSTWU module computes the column's minimum reflux ratio, minimum number of stages, and feed stage location, providing a baseline for subsequent design refinements.

Once the prefractionator's preliminary parameters are established, the next step involves the rigorous modeling of the main column sections. These are treated as two interconnected distillation columns, one handling the separation of lighter components and the other focusing on heavier components. Internal flows—vapor and liquid—are iteratively adjusted to maintain energy balance between the prefractionator and the main column. This iterative process ensures that the thermal coupling is accurately represented and that the energy savings potential of the DWC is fully realized. The method incorporates a balancing step where the vapor and liquid flows from the prefractionator are matched with those entering the main column sections. This alignment is critical to achieving both operational stability and efficient energy use (Buitimea-Cerón et al., 2021).

Incorporation of advanced simulation and sensitivity analysis

A defining feature of the T&S methodology is its flexibility in integrating with advanced simulation tools. Modern implementations often rely on process simulators like Aspen Plus or HYSYS to refine the initial estimates. Rigorous simulations using RadFrac or equivalent modules adjust for deviations introduced by simplifying assumptions, such as constant molar flows and relative volatilities. Additionally, sensitivity analyses are employed to evaluate the impact of key design variables—such as reflux ratio, feed composition, and split ratios—on column performance. These analyses guide iterative adjustments to ensure that the final design is robust and optimized.

Challenges and modern adaptations

While the T&S methodology provides a robust framework for DWC design, certain challenges persist. The reliance on Gilliland and Kirkbride correlations, for instance, may lead to inaccuracies when transitioning from shortcut to rigorous simulations. These discrepancies often arise from nonideal behaviors or operational constraints that are not captured in the simplified models. To address this, contemporary adaptations of the T&S method integrate optimization algorithms and dynamic control strategies, which refine the thermal coupling and stage distribution in real-time. The use of optimization tools further ensures that the column operates near its energy-efficient limits while maintaining the required separation specifications.

The T&S methodology remains a cornerstone in DWC design, offering a structured and adaptable approach for tackling the complexity of thermally coupled systems. It is particularly useful in industrial applications involving multi-component separations, where its iterative nature allows for fine-tuning of interconnection flows and energy distribution. The methodology has also been successfully adapted for azeotropic and reactive separations, highlighting its versatility. By providing a reliable foundation for column configuration and operation, the T&S method enables designers to achieve substantial energy savings and separation efficiencies.

V_{min} methodology for determining the minimum steam consumption

The V_{min} Diagram Method is a graphical technique developed by Halvorsen and Skogestad to determine the minimum vapor flow required for a distillation process. It is particularly effective for thermally coupled systems like the Petlyuk column and DWCs. This methodology leverages the Underwood equations, which assume constant molar flows, constant relative volatilities, and an infinite number of stages. These simplifications make the method highly useful for estimating the energy requirements and internal flow dynamics of DWCs during the early stages of design. The V_{min} Diagram is a practical tool that provides insight into the energy demands for specific separations and aids in identifying the most challenging splits within the column (Ränger et al., 2018).

Fundamentals of the V_{min} diagram

The V_{min} Diagram Method calculates the minimum vapor flow required for each binary split in a multicomponent mixture. For ternary systems, the primary splits considered are *A*/*BC*, *AB*/*C*, and *B*/*AC*, where *A*, *B*, and *C* are the components arranged in order of decreasing volatility. The diagram plots the normalized top vapor flow rate (VT/F) as a function of the product split (D/F), with the peak of the curve representing the most demanding separation. This peak corresponds to the minimum vapor flow required to achieve the desired separations with the least energy input (Engelien & Skogestad, 2005).

The methodology uses the following equations to determine the minimum vapor flow for a Petlyuk column handling ternary mixtures:

$$V_{\text{min,Petlyuk}} = \max\left(\frac{\alpha_A z_A}{\alpha_A - \theta_A} + \frac{\alpha_A z_A}{\alpha_A - \theta_B} + \frac{\alpha_B z_B}{\alpha_B - \theta_B}\right) F \qquad (7.2)$$

In this equation:

F is the total feed flow rate.
z_A and z_B are the feed compositions of components A and B.
α_A and α_B are the relative volatilities of components A and B.
θ_A and θ_B are underwood roots determined by the desired product split.

The relationship between feed thermal quality (q) and vapor flow is expressed as:

$$V_{T\text{min,Petlyuk}} = V_{B\text{min,Petlyuk}} + (1 - q) \quad F$$

This equation ensures that the interplay between feed condition and energy distribution is captured, facilitating accurate estimations for systems with varying feed thermal qualities (Fig. 7.1) (Dejanović et al., 2014).

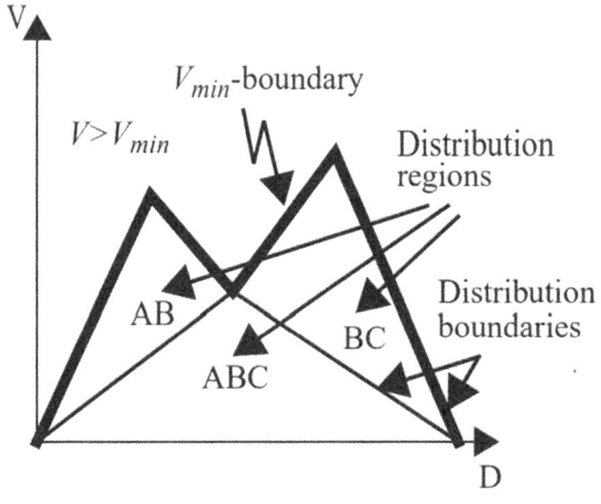

FIGURE 7.1 V_{min} diagram for a ternary mixture.

Practical implementation in process simulators

The V_{min} Diagram Method has been seamlessly integrated into process simulators like Aspen Plus and HYSYS, enabling users to transition from conceptual designs to rigorous simulations. Shortcut models, such as those generated using the DSTWU block, estimate minimum stages (N_{min}) and reflux ratios. These results form the basis for constructing the V_{min} diagram, which visualizes the energy distribution for different product splits. In rigorous simulations, modules like RadFrac are used to refine the vapor and liquid flow rates, ensuring that product purity and energy efficiency are simultaneously achieved.

Typically, the infinite-stage assumption in the V_{min} Diagram Method is approximated by using four times the minimum number of stages ($4N_{min}$). This scaling factor provides a practical balance between idealized assumptions and real-world conditions. Parameters such as boil-up ratios, side draw flows, and reflux ratios are adjusted iteratively during these simulations to identify the optimal configuration for the DWC.

Design insights and limitations

The V_{min} Diagram is particularly effective in identifying the most energy-intensive splits in a distillation process, making it invaluable for processes involving multicomponent mixtures. By focusing on the vapor flow requirements, engineers can prioritize design modifications that address the most significant energy demands, such as adjusting stage distribution or optimizing thermal coupling. This insight helps minimize the overall energy consumption of the column and ensures its operational feasibility.

However, the method does have limitations. While it excels in estimating energy requirements, it does not provide detailed structural information about the column, such as the number of trays in each section or the precise placement of interconnection flows. Heuristic rules are often employed to bridge this gap, such as assigning the main column stages twice the combined minimum stages for *A*/*BC* and *AB*/*C* separations, while the prefractionator stages are distributed based on the stripping and rectifying requirements of these splits (Halvorsen & Skogestad, 2003).

Broader implications and use cases

The V_{min} Diagram Method has found widespread application in industrial settings, particularly for the initial design and optimization of thermally coupled columns. Its graphical representation provides an intuitive understanding of the trade-offs involved in separation tasks, aiding in decision-making during early design stages. Furthermore, its integration with other methodologies, such as the Triantafyllou & Smith approach or sensitivity analyses, enhances its applicability to complex, nonideal mixtures and dynamic operational conditions.

Thus, The V_{min} Diagram Method remains a powerful tool for the preliminary design of DWC s and other thermally coupled distillation systems. By providing a clear visualization of energy requirements and vapor flow distributions, it simplifies the complex task of designing energy-efficient and high-performance columns. While its assumptions limit its standalone applicability, its integration with rigorous simulation tools and complementary design strategies ensures that it remains a vital resource in modern distillation technology (Skiborowski et al., 2022, p. 431–481).

Sotudeh & Shahraki methodology

The Sotudeh & Shahraki (S&S) methodology provides a systematic and analytical framework for designing DWCs. This method is built upon the Underwood equations but diverges from traditional techniques by avoiding reliance on the Fenske equation. This distinction makes the S&S methodology particularly effective for scenarios where liquid and vapor compositions are not equal at interconnecting stages, a common occurrence in multicomponent and azeotropic separations. Its adaptability to both ternary and multicomponent mixtures has made it a valuable tool in the field of distillation design (Sotudeh & Shahraki, 2007).

The S&S methodology employs a structured two-stage design process:

Calculation of operational variables: The minimum vapor flow rates (V_{min}) are calculated using principles derived from the V_{min} methodology. The recovery of the middle-boiling component at the top of the prefractionator is used as a key parameter, ensuring a clear separation between light and heavy components. This step establishes the basis for optimizing energy efficiency and flow distributions.

Determination of the Number of Stages: The number of theoretical stages in the prefractionator and main column sections is determined using a Petlyuk system configuration. This involves solving the Underwood equations for both sections, ensuring that the internal flows align with the desired product specifications and operational constraints (Buitimea-Cerón et al., 2021).

The methodology incorporates a series of critical equations for determining minimum vapor flow, reflux ratios, liquid splits, and the number of theoretical stages:

- Minimum vapor flow calculation (see Eq. 7.2)
- Reflux ratio (RR_{min}):

$$RR_{min} = \frac{V_{Petlyuk,min}}{D_2} - 1$$

where is the distillate flow rate

- Liquid split parameter (r): The liquid split r is a critical parameter for balancing the prefractionator and main column flows. It is constrained by:

$$\max \left[\frac{L_{\min,2}}{L_2}, \frac{L_{\min,3}+S}{L_2}\right] < r < \frac{L_2 - L_{\min,1} + qf}{L_2}$$

where q represents the thermal quality of the feed.

- Number of theoretical stages: The number of stages in the prefractionator and main column sections is determined using the Underwood roots and component compositions:

$$N_3 = log \frac{\left(\frac{\sum \frac{\alpha_i x_i D2}{\alpha_i \phi_2'}}{\sum \frac{\alpha_i x_i D2}{\alpha_i \phi_1'}}\right)}{\log\left(\frac{\phi_2'}{\phi_1'}\right)}$$

$$N_4 = log \frac{\left(\frac{\sum \frac{\alpha_i x_i L}{\alpha_i \psi_2'}}{\sum \frac{\alpha_i x_i L}{\alpha_i \psi_1'}}\right)}{\log\left(\frac{\psi_2'}{\psi_1'}\right)}$$

Practical considerations

The S&S methodology ensures that DWCs are designed with energy efficiency and operational feasibility in mind. The focus on minimizing vapor flow and balancing liquid splits reduces energy consumption while maintaining high separation efficiency. However, the methodology requires precise selection of parameters, particularly for the liquid split (r), as deviations can significantly impact column performance.

The S&S methodology has been successfully applied to various industrial separation tasks, including those involving azeotropic and highly nonideal mixtures. Its reliance on the Underwood equations allows for a detailed and analytical approach to column design. Nevertheless, it assumes ideal behavior in some aspects, which may limit its accuracy for highly nonideal or reactive mixtures. Incorporating rigorous simulations alongside the S&S methodology can address these limitations, ensuring robust designs under real-world conditions.

Thus, the Sotudeh & Shahraki methodology represents a robust and versatile approach to designing DWCs. By focusing on minimum vapor flows, liquid splits, and precise stage calculations, it offers a comprehensive framework for achieving energy-efficient and operationally stable designs. Its adaptability to multicomponent mixtures and analytical rigor make it an indispensable tool for engineers in the field of distillation technology (Buitimea-Cerón et al., 2021; Sotudeh & Shahraki, 2007).

Case study: Comparison of energy consumptions in the separation of a ternary mixture considering different design methodologies

To evaluate the effectiveness of various methodologies for designing DWCs, a comprehensive analysis of case studies involving ternary mixtures was conducted. These mixtures were selected based on their Ease of Separation Index (ESI), a parameter defined as, where α_{AB} and α_{BC} are the relative volatilities of the *A*/*B* and *B*/*C* splits, respectively. This index categorizes the difficulty of separations, with values below 1 indicating a challenging *A*/*B* split and values above 1 highlighting a more complex *B*/*C* separation. By considering a diverse range of ESI values, this section provides insight into the robustness and applicability of DWC design methodologies under varying separation conditions (Buitimea-Cerón et al., 2021).

Thus, five representative case studies were examined, each characterized by distinct relative volatilities, feed compositions, and thermodynamic models. These mixtures range from those with relatively straightforward separations to those involving highly complex interactions. The selected cases ensure a balanced assessment of separation challenges:

- BTX Mixture (Benzene, Toluene, Xylene):
  With an ESI of 0.83, this mixture represents moderately challenging separations. The feed consists of 30% benzene, 30% toluene, and 40% xylene, modeled using the Chao-Seader thermodynamic framework. Operating at 358 K, this case highlights intermediate difficulties in splitting the *A*/*B* and *B*/*C* components.
- PHH Mixture (n-Pentane, n-Hexane, n-Heptane):
  This mixture, with an ESI of 1.04, represents balanced separation challenges. A feed flow of 45.4 kmol/h at 2.04 atm and equal compositions of 33% for each component is modeled using the RK-Soave thermodynamic equation. It provides a standard case for assessing the methodologies' baseline performance.
- BPP Mixture (n-Butane, i-Pentane, n-Pentane):
  Characterized by an ESI of 1.86, this mixture features a relatively easier *A*/*B* split compared to the *B*/*C* separation. Operating at a pressure of 6.17 atm, this case examines the performance of DWC designs for higher

ESI values, reflecting minimal separation complexity in the initial fractionation stages.

- BBH Mixture (i-Butane, n-Butane, n-Hexane):
  With an ESI of 0.18, this mixture poses significant challenges for the *A/B* split while simplifying the *B/C* separation. Modeled at 7.7 atm using the RK-Soave model, this case underscores the importance of robust methodologies for difficult separations.
- PPH Mixture (i-Pentane, n-Pentane, n-Hexane):
  This mixture has an ESI of 0.47, reflecting moderately challenging *A/B* splits. Operating at 2.55 atm with feed compositions of 33% for each component, it serves as a test case for moderate separation complexity using the RK-Soave thermodynamic framework.

All simulations were performed using Aspen Plus with rigorous RadFrac column modeling. Design specifications were set to achieve product purities of 99 mol%, with adjustable reflux ratios and flow rates. The pressure of the columns was optimized to ensure efficient heat transfer, leveraging cooling water at 25°C with a 4°C rise. The distillate temperature was maintained at 39°C to ensure compatibility with the cooling system. These standardized parameters provide consistency across the case studies, enabling direct comparisons of methodology performance under varying separation challenges. This analysis offers a practical foundation for assessing the effectiveness of different design methodologies for DWCs. The diversity of case studies ensures that insights gained are broadly applicable, supporting the development of robust and energy-efficient designs. The following sections will explore the outcomes and implications of these simulations, emphasizing the role of case-specific parameters in guiding optimal design strategies.

Results and discussion

Each design methodology focuses on generating feasible parameters to initialize the column configurations, followed by the application of optimization techniques when necessary. This section presents a comparative analysis of the key parameters obtained from the DWC designs using each methodology. To ensure a consistent comparison, product purity specifications were set at 99 mol% for all streams. While product recoveries were not explicitly defined as constraints, it is worth noting that, across all case studies, the recoveries exceed 0.99 for the light and heavy components (*A* and *C*) and are greater than 0.97 for the intermediate component (*B*) (Tables 7.1–7.5).

The performance of different methodologies for designing DWCs varies significantly based on the properties of the mixtures, particularly the ease of separation index (ESI). The ESI serves as a critical indicator, defining the relative difficulty of separating the components within a mixture. As such, methodologies must be evaluated based on their ability to adapt to various separation challenges and produce efficient and feasible designs.

TABLE 7.1 Results for the benzene, toluene, and o-xylene mixture.

Parameter	V_{min} (prefrac)	V_{min} (main column)	T&S (prefrac)	T&S (main column)	S&S (prefrac)	S&S (main column)
Stages	24	48	20	44	17	35
Feed stage	15	15	12	14	9	10
Side stage	24		20		19	
Reflux ratio	2.94		2.98		3.76	
L-p (kmol/h)	817		901		1013	
V-p (kmol/h)	2297		2321		2648	
Diameter (m)	9.74		9.86		10.78	
QCond (kW)	38,353		39,025		46,456	
QReb (kW)	41,215		41,661		48,740	
TAC ($/year)	12,390,000		12,433,000		14,177,296	

TABLE 7.2 Results for the n-Pentane, n-Hexane, and n-Heptane (PHH) mixture.

Parameter	V_{min} (prefrac)	V_{min} (main column)	T&S (prefrac)	T&S (main column)	S&S (prefrac)	S&S (main column)
Stages	26	52	22	46	19	38
Feed stage	16	15, 41	14	13, 35	10	9, 29
Side stage	24		18		18	
Reflux ratio	3.36		3.95		9.78	
L-p (kmol/h)	11.9		12.4		15	
V-p (kmol/h)	27.3		28.9		37	
Diameter (m)	0.79		0.83		1.15	
QCond (kW)	467		530		1161	
QReb (kW)	469		531		1163	
TAC ($/year)	255,630		272,240		485,648	

TABLE 7.3 Results for the n-Butane, i-Pentane, and n-Pentane (BPP) mixture.

Parameter	V_{min} (prefrac)	V_{min} (main column)	T&S (prefrac)	T&S (main column)	S&S (prefrac)	S&S (main column)
Stages	61	123	69	124	64	119
Feed stage	35	15, 76	38	12, 81	32	10, 74
Side stage	26		18		24	
Reflux ratio	14.2		13.49		43.41	
L-p (kmol/h)	36.5		32.8		53.2	
V-p (kmol/h)	46.7		44.3		72.1	
Diameter (m)	1		0.98		1.71	
QC (kW)	1282		1219		3761	
QR (kW)	1268		1206		3747	
TAC ($/year)	631,260		607,810		1,500,009	

TABLE 7.4 Results for the i-butane, n-butane, and n-hexane (BBH) mixture.

Parameter	V_{min} (prefrac)	V_{min} (main column)	T&S (prefrac)	T&S (main column)	S&S (prefrac)	S&S (main column)
Stages	40	84	34	73	42	79
Feed stage	20	38, 78	16	36, 70	23	34, 76
Side stage	72		66		71	
Reflux ratio	43.78		14.3		13.42	
L-p (kmol/h)	11.5		10		10.1	
V-p (kmol/h)	30.4		28		37.8	
Diameter (m)	1.54		0.92		0.91	
QC (kW)	3359		1154		1090	
QR (kW)	3391		1185		1120	
TAC ($/year)	1,344,900		551,080		537,783	

TABLE 7.5 Results for i-pentane, n-pentane, and n-hexane (PPH) mixture.

Parameter	V_{min} (prefrac)	V_{min} (main column)	T&S (prefrac)	T&S (main column)	S&S (prefrac)	S&S (main column)
Stages	51	115	40	96	48	99
Feed stage	32, 55	106	24, 52	92	26, 37	85
Side stage	90		83		77	
Reflux ratio	32.32		12.54		17.89	
L-p (kmol/h)	24.3		23.9		23.6	
V-p (kmol/h)	43.9		43.5		50.4	
Diameter (m)	1.77		1.17		1.37	
QC (kW)	3340		1360		1911	
QR (kW)	3331		1352		1903	
TAC ($/year)	1,386,500		649,290		836,837	

For mixtures with low ESI values (below 0.5), certain methodologies often lead to designs characterized by higher energy consumption and increased total annual costs (TAC). These outcomes can be attributed to the difficulty of accurately estimating interconnecting flows and structural parameters, which are critical to minimizing energy use and ensuring operational feasibility. In such cases, the overestimation of reflux ratios and stage requirements further exacerbates the inefficiencies, resulting in higher utility consumption and less favorable economic performance. For intermediate ESI values (around 0.83), methodologies such as V_{min} and T&S demonstrate greater reliability. These approaches effectively balance the separation effort between components, leading to energy-efficient designs with more reasonable reflux ratios and stage distributions. This performance consistency can be attributed to the adaptability of these methods in handling mixtures where the separation difficulty is evenly distributed across the components.

In contrast, for mixtures with high ESI values (e.g., 1.86), methodologies that rely heavily on simplified assumptions, such as basic Underwood calculations, face significant limitations. These approaches often struggle to predict the correct number of stages or optimize reflux ratios, which leads to suboptimal designs with higher energy demands. The difficulty in capturing the complex interactions between components and the increased reliance on heuristics to estimate liquid splits and vapor flows further reduce the effectiveness of these methodologies in such scenarios. Additionally, these methods may fail to leverage the thermodynamic advantages of DWCs, such as the potential for energy recovery through thermal coupling. Another critical consideration in comparing methodologies is their sensitivity to feed composition and product specifications. While most methodologies aim to achieve product purities of 99 mol%, the distribution of product recoveries (particularly for intermediate components) can vary significantly depending on the design strategy. Accurate estimation of interconnection flows and precise stage allocations are essential to achieving high recoveries and minimizing energy use.

The differences in performance across methodologies highlight the importance of understanding the limitations and assumptions underlying each approach. Methodologies that rely on simplified stage and flow predictions may be suitable for cases with balanced separation challenges but are less effective for highly asymmetric mixtures. Conversely, approaches that incorporate iterative adjustments to liquid splits and vapor flows may better accommodate the complexities of multicomponent systems, albeit with increased computational demands. Ultimately, the choice of design methodology should be guided by the properties of the mixture, including its ESI, feed composition, and thermodynamic characteristics. Designers must carefully evaluate the trade-offs between accuracy, computational complexity, and energy efficiency to select the most appropriate approach for the given separation task. This comparative analysis underscores the need for a nuanced understanding of each methodology's capabilities and limitations, particularly as DWCs continue to play a vital role in energy-efficient distillation processes.

Dividing-wall columns (DWCs) have emerged as a promising technology for sustainable separation processes. Their ability to significantly reduce energy consumption compared to conventional separation sequences highlights their potential in the context of process intensification. The proper selection of design methodologies plays a crucial role in achieving near-optimal DWC configurations. This ensures not only energy savings but also operational and economic efficiency. While significant progress has been made in developing and applying various design methodologies, several challenges and research opportunities remain. For instance, the impact of feed composition on the performance of each design methodology has not been extensively analyzed. Future work should aim to address this gap by evaluating how variations in feed composition influence the structural and operational parameters of DWC designs. Moreover, additional case studies incorporating a broader range of mixtures and separation difficulties could provide valuable insights into the strengths and limitations of each methodology, ultimately leading to more generalized design guidelines. Another intriguing research avenue is the application of DWC methodologies to azeotropic mixtures. The current methodologies are primarily tailored for separating non-azeotropic mixtures. Expanding their scope to handle azeotropic systems would require methodological adjustments, such as incorporating advanced thermodynamic models and refining initialization procedures. Addressing this limitation could open new possibilities for the application of DWCs in more complex separation scenarios.

A comparative summary of the investigated methodologies reveals their unique strengths and limitations:

- V_{min} methodology: This method is straightforward to implement as it relies solely on the Underwood equations. It adapts well to process simulators and produces near-optimal designs for mixtures with an ease of separation index (ESI) > 0.8. However, it lacks the ability to provide detailed tray arrangements and performs poorly for mixtures with low ESI values, yielding large reflux ratios and higher total annual costs (TAC).
- Triantafyllou & Smith (T&S) Methodology: The T&S methodology is versatile and delivers reliable designs regardless of the mixture's ESI. It is easily integrated with process simulators. However, its complexity surpasses that of the V_{min} methodology because it incorporates additional equations (e.g., FUGK equations) alongside the Underwood equations.
- Sotudeh & Shahraki (S&S) Methodology: This methodology provides equations for determining the column's topological arrangement, making it advantageous for initial design steps. However, the economic performance of the resulting designs is highly sensitive to the assumptions made, which can lead to suboptimal results in certain cases.

These insights underscore the importance of selecting an appropriate design methodology based on the specific characteristics of the mixture to be separated. Methodologies should be chosen not only for their ease of implementation but also for their ability to adapt to the complexity of the separation process and deliver energy-efficient designs.

General conclusions

The study of DWCs presented in this chapter underscores their potential as a transformative technology in distillation processes. By integrating insights from theoretical models, simulation parameters, and case studies, the chapter provides a holistic understanding of DWC design and optimization. A central theme is the recognition that the performance of DWCs is intricately linked to the proper selection of design methodologies, tailored to the specific properties of the mixtures being separated. The analysis of design variables and degrees of freedom reveals the complexity inherent in DWC configurations, particularly the role of interconnection flows. These flows are critical not only for reducing energy consumption but also for achieving precise component separations and optimizing column performance. The thermodynamic and operational implications of these flows highlight the need for methodologies that balance energy efficiency with operational feasibility.

A comprehensive evaluation of established design methodologies, including V_{min}, Triantafyllou & Smith, and Sotudeh & Shahraki, demonstrates their varying strengths and limitations. The V_{min} methodology, while simple and effective for intermediate ease of separation indices (ESI), struggles with mixtures requiring high or low separation efforts. Conversely, methodologies like T&S offer greater versatility, delivering reliable designs across a broader spectrum of separation challenges. The chapter's comparative analysis through case studies showcases how differences in reflux ratios, stage distributions, and energy requirements can significantly impact the total annual cost (TAC) of DWC designs. The simulation studies emphasize the importance of robust initialization strategies and the refinement of design parameters through rigorous modeling. This approach ensures that designs not only meet purity specifications but also achieve optimal recoveries for all components. The findings highlight the necessity of combining theoretical frameworks with practical adjustments to address real-world operational complexities.

In considering future directions, the chapter identifies opportunities for advancing DWC methodologies, particularly in the context of azeotropic mixtures and more complex separation scenarios. The integration of advanced thermodynamic models and the exploration of novel configurations hold promise for further enhancing the applicability and efficiency of DWCs. Overall, this chapter establishes a foundation for understanding the intricacies of DWC design and optimization. The findings emphasize the critical role of tailored methodologies in harnessing the full potential of DWCs, paving the way for more energy-efficient and economically viable solutions in modern distillation practices.

References

Buck, C., Hiller, C., & Fieg, G. (2011). Applying model predictive control to dividing wall columns. *Chemical Engineering and Technology, 34*(5), 663–672. https://doi.org/10.1002/ceat.201000487

Buitimea-Cerón, G. A., Hahn, J., Medina-Herrera, N., Jiménez-Gutiérrez, A., Loredo-Medrano, J. A., & Tututi-Avila, S. (2021). Dividing-wall column design: Analysis of methodologies tailored to process simulators. *Processes, 9*(7). https://doi.org/10.3390/pr9071189

Christiansen, A. C., Skogestad, S., & Lien, K. (1997). Complex distillation arrangements: Extending the Petlyuk ideas. *Computers and Chemical Engineering, 21*(1), S237. https://doi.org/10.1016/S0098-1354(97)87508-4

Dejanović, I., Halvorsen, I. J., Skogestad, S., Jansen, H., & Olujić, Ž. (2014). Hydraulic design, technical challenges and comparison of alternative configurations of a four-product dividing wall column. *Chemical Engineering and Processing: Process Intensification, 84*, 71–81. https://doi.org/10.1016/j.cep.2014.03.009

Engelien, H. K., & Skogestad, S. (2005). Minimum energy diagrams for multieffect distillation arrangements. *AIChE Journal, 51*(6), 1714–1725. https://doi.org/10.1002/aic.10453

Fang, J., Zhao, H., Qi, J., Li, C., Qi, J., & Guo, J. (2015). Energy conserving effects of dividing wall column. *Chinese Journal of Chemical Engineering, 23*(6), 934–940. https://doi.org/10.1016/j.cjche.2014.08.009

Flores, O. A., Cárdenas, J. C., Hernández, S., & Rico-Ramírez, V. (2003). Thermodynamic analysis of thermally coupled distillation sequences. *Industrial and Engineering Chemistry Research, 42*(23), 5940–5945. https://doi.org/10.1021/ie034011n

Halvorsen, I. J., & Skogestad, S. (2003). Minimum energy consumption in multicomponent distillation. 3. More than three products and generalized Petlyuk arrangements. *Industrial and Engineering Chemistry Research, 42*(3), 616–629. https://doi.org/10.1021/ie0108651

Ho, Y. C., Ward, J. D., & Yu, C. C. (2011). Quantifying potential energy savings of divided wall columns based on degree of remixing. *Industrial and Engineering Chemistry Research, 50*(3), 1473–1487. https://doi.org/10.1021/ie901907e

Ignat, R. M., & Kiss, A. A. (2013). Optimal design, dynamics and control of a reactive DWC for biodiesel production. *Chemical Engineering Research and Design, 91*(9), 1760–1767. https://doi.org/10.1016/j.cherd.2013.02.009

Lestak, F., & Collins, C. (1997). Advanced distillation saves energy and capital. *Chemical Engineering, 104*(7).

Ränger, L. M., Preißinger, U., & Grützner, T. (2018). Robust initialization of rigorous process simulations of multiple dividing wall columns via V_{min} diagrams. *ChemEngineering, 2*(2), 1–19. https://doi.org/10.3390/chemengineering2020025

Sangal, V. K., Kumar, V., & Mishra, I. M. (2012). Divided wall distillation column: Rationalization of degree of freedom analysis. *Theoretical Foundations of Chemical Engineering, 46*(4), 319–328. https://doi.org/10.1134/s0040579512040148

Skiborowski, M., Kruber, K. F., & Waltermann, T. (2022). *Sustainable distillation processes sustainable separation engineering: Materials, techniques and process development.* Wiley. https://doi.org/10.1002/9781119740117.ch12

Sotudeh, N., & Shahraki, B. H. (2007). A method for the design of divided wall columns. *Chemical Engineering and Technology, 30*(9), 1284–1291. https://doi.org/10.1002/ceat.200700175

Suphanit, B., Bischert, A., & Narataruksa, P. (2007). Exergy loss analysis of heat transfer across the wall of the dividing-wall distillation column. *Energy, 32*(11), 2121–2134. https://doi.org/10.1016/j.energy.2007.04.006

Chapter 8

Complex divided wall column arrangements

Motivation and theoretical aspects of DWCs with more than one wall

Distillation, a cornerstone of industrial chemical processes, remains the most energy-intensive separation method worldwide. This reality has driven extensive research and innovation within thermal process engineering, aiming to enhance its energy efficiency and economic viability (Górak & Schoenmakers, 2014, p. 1–445). Among the advancements, the dividing wall column (DWC) has emerged as a pivotal technology, offering significant savings in both operational and investment costs (Fig. 8.1). By integrating dividing walls into a single shell, DWCs enable the efficient separation of ternary or higher-order mixtures with reduced energy consumption and capital expenditure (Dejanović et al., 2011). Traditional distillation sequences for ternary mixtures typically involve multiple columns arranged in direct or indirect splits, each associated with inherent inefficiencies due to remixing of intermediate boiling components. This remixing generates entropy, decreasing thermodynamic efficiency and increasing energy demand (Hernández, 2008). DWCs overcome this limitation by creating distinct separation sections within a single shell, minimizing internal mixing and remixing of liquid and vapor streams. This structural advantage translates into energy savings of up to 30% compared to conventional configurations, along with a similar reduction in investment costs (Soraya Rawlings et al., 2019).

The DWC is thermodynamically equivalent to the Petlyuk configuration but achieves the same separation within one column shell. This design prevents lateral mixing of streams through the dividing wall and allows precise control over vapor and liquid splits. However, this enhanced performance comes with increased complexity. A conventional two-column sequence typically has 10 degrees of freedom, while a three-product DWC introduces 12 degrees of freedom, encompassing operating pressure, theoretical stages in six sections, product specifications, and splits at the dividing wall (Yildirim et al., 2011).

Thermally Coupled Distillation Columns. https://doi.org/10.1016/B978-0-443-33114-5.00004-1

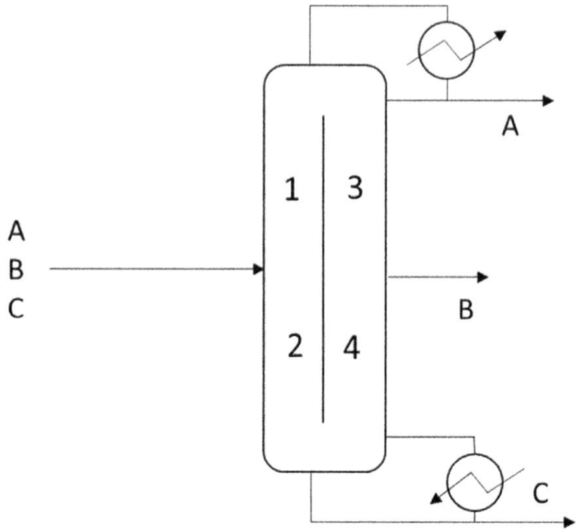

FIGURE 8.1 DWC for a ternary mixture separation.

Extending the DWC concept: Multiple dividing walls

The extension of the DWC concept to configurations with multiple dividing walls, referred to as multi-DWCs (mDWCs), represents a transformative advancement in distillation technology. While standard DWCs have proven highly effective for ternary mixtures, mDWCs expand the scope of application to include mixtures with four or more components. This development addresses the limitations of conventional distillation sequences, offering enhanced energy efficiency and reduced operational complexity for multicomponent separations. At the core of mDWC design lies the thermodynamic principle of minimizing entropy generation caused by mixing and remixing of components during separation. Traditional column sequences, such as the direct split or indirect split configurations, often suffer from these inefficiencies, as intermediate boiling components are subjected to unnecessary remixing within the separation process (Górak & Schoenmakers, 2014, p. 1–445). In contrast, mDWCs employ additional dividing walls to create independent separation zones, preventing lateral mixing and enabling simultaneous separation of multiple components in a single column shell (Fig. 8.2).

The thermodynamic equivalence between mDWCs and the Petlyuk configuration underpins their design. The Petlyuk arrangement achieves optimal energy efficiency by performing sharp splits between adjacent boiling components across interconnected columns. However, mDWCs integrate these splits within a single shell, significantly reducing capital and operational expenses. For instance, in a four-product mDWC designed for a mixture ABCD, the column is segmented into prefractionator, middle, and main column

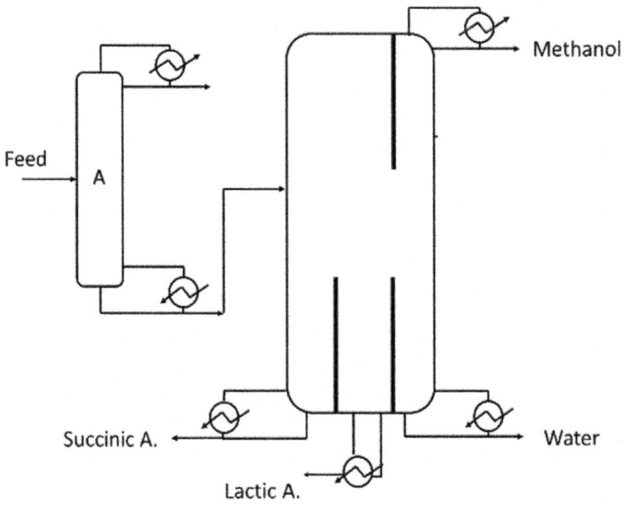

FIGURE 8.2 Example of multiple dividing wall columns.

sections, each responsible for isolating specific components. The absence of mixing between these zones results in energy savings of up to 55% compared to conventional distillation sequences (Dejanović et al., 2010). The introduction of multiple dividing walls offers additional flexibility for separating zeotropic mixtures with varying boiling points. By tailoring the number and placement of walls, mDWCs can achieve nearly any desired separation, making them suitable for a broad range of industrial applications. This flexibility is particularly valuable in processes where feed composition and product specifications may vary, as the column can be reconfigured to adapt to changing conditions without requiring significant modifications to its structure (Hernández & Jiménez, 1999).

Despite their potential, mDWCs present unique challenges in both design and operation. The addition of multiple dividing walls increases the number of degrees of freedom significantly, from 12 in a standard three-product DWC to 23 in a four-product mDWC. These degrees of freedom include operating pressure, theoretical stages across 12 sections, vapor and liquid splits at each dividing wall, and product specifications. Managing this complexity requires advanced modeling tools and optimization techniques to identify configurations that balance efficiency, cost, and operational feasibility. One of the key challenges in mDWC design is ensuring hydraulic stability across multiple separation zones. Uneven vapor or liquid flow distribution can lead to inefficiencies or operational instabilities, particularly when dealing with mixtures that exhibit nonideal behavior. For example, vapor splits are influenced by pressure drop variations across the column, which depend on factors such as feed composition, flow rates, and internal column design. Advanced computational models, such as computational fluid dynamics (CFD), have been employed to simulate these phenomena and optimize the internal design of mDWCs (Soraya Rawlings et al., 2019).

Heat transfer across dividing walls introduces another layer of complexity. While the walls are intended to isolate separation zones, unintended thermal interactions can occur, leading to deviations from the desired separation profile. Studies have shown that these effects can be mitigated by using specialized column internals, such as structured packing or trays, designed to minimize thermal bridging while maintaining high mass transfer efficiency. Additionally, precise control of liquid and vapor splits is essential to achieve optimal separation. While liquid splits can be adjusted using flow control devices, vapor splits are more challenging to manipulate and often rely on passive mechanisms, such as differential pressure regulation, to maintain balance (Rong, 2011).

Industrial applications of mDWCs

The versatility of mDWCs makes them particularly appealing for industries that require the separation of complex feedstocks into high-purity products. Several key applications include:

- Petrochemical industry:
  The separation of complex hydrocarbon mixtures, such as those derived from crude oil or natural gas, often involves numerous components with close boiling points. mDWCs provide a compact and energy-efficient solution for these separations, reducing the need for multiple distillation columns and auxiliary equipment. For instance, an mDWC can simultaneously separate light hydrocarbons, such as propane and butane, from heavier fractions, improving overall process efficiency (Segovia-Hernández et al., 2021).
- Biofuel production:
  The production of bioethanol and biobutanol from fermentation processes often results in feed mixtures containing water, alcohols, and other impurities. Traditional distillation setups struggle with azeotropic behavior and high energy demands for these separations. mDWCs offer a promising alternative by enabling the simultaneous recovery of ethanol, butanol, and water in a single column, reducing energy consumption and improving product yields (Errico et al., 2017).
- Fine chemical and pharmaceutical sectors:
  The synthesis of high-value chemicals often involves complex reaction mixtures that require precise separation of multiple components. mDWCs are particularly suited for these applications due to their ability to handle diverse feed compositions and stringent product purity requirements. For example, the separation of chiral compounds or active pharmaceutical ingredients can be achieved more efficiently using mDWC technology (Segovia-Hernández et al., 2021).

- Sustainability-oriented applications:
  As industries increasingly prioritize sustainability, mDWCs align with efforts to reduce carbon footprints and energy consumption. By integrating advanced control systems and leveraging renewable energy sources, mDWCs can be adapted to meet the demands of green chemistry and circular economy initiatives. This potential positions them as a key technology for achieving sustainable process intensification (PI) (Segovia-Hernández et al., 2021).

Given the inherent complexity of mDWCs, research has focused on simplifying their design while retaining their performance advantages. One notable approach involves reducing the number of dividing walls without compromising separation efficiency. For example, the 2–2–4-a configuration eliminates one dividing wall between the middle and main columns, reducing the degrees of freedom from 23 to 19. This simplification not only eases the design process but also enhances hydraulic stability and operational flexibility. Innovations in column internals have also contributed to the feasibility of mDWCs. Enhanced packing materials, such as high-efficiency structured packing, have been developed to improve mass transfer rates and reduce pressure drop. Additionally, modular designs that allow for easy reconfiguration of column sections have been proposed to accommodate changing feed compositions and product demands (Sotudeh & Shahraki, 2007).

Challenges and opportunities (Sotudeh & Shahraki, 2007)

While the benefits of mDWCs are clear, their implementation comes with unique challenges that must be addressed to unlock their full potential. These challenges span operational, design, and practical considerations, each of which requires targeted solutions to facilitate broader industrial adoption. One of the most significant operational challenges in mDWCs is the control of vapor and liquid splits at each dividing wall. Unlike liquid splits, which can be adjusted with relative ease using flow control devices, vapor splits are often governed by the pressure drop across the column. Achieving a balanced vapor split is critical for maintaining product purity and column stability. However, the dynamic nature of vapor flow, influenced by factors such as feed composition and internal hydraulics, complicates this process. Advanced control strategies, such as model predictive control (MPC), have been proposed to manage these complexities. By integrating real-time data with predictive algorithms, MPC systems can dynamically adjust operating parameters to maintain optimal performance (Kiss, 2013). Nonetheless, these systems require accurate modeling and robust computational resources, which may limit their application in smaller-scale operations. The design of mDWCs also poses significant challenges, particularly regarding the placement and configuration of dividing walls. The number and position of these walls directly influence the column's separation efficiency and energy consumption. For instance, the 2–2–4-a configuration simplifies the traditional 2–3–4 design by eliminating one dividing wall between the middle and main columns. This modification reduces the degrees of freedom from 23 to 19, easing the design process while maintaining comparable energy efficiency. Additionally, heat transfer across dividing walls introduces uncertainties that complicate predictive modeling. This phenomenon can create hot or cold spots within the column, affecting the distribution of vapor and liquid flows. Experimental studies have shown that these effects can be mitigated by optimizing the design of column internals, such as structured packing or trays, to ensure uniform flow distribution (Lukač et al., 2019).

In practical applications, mDWCs face limitations related to column height and mechanical stability. The increased number of separation sections often results in taller columns, which may be impractical for existing facilities with height restrictions. Moreover, the added structural complexity increases manufacturing and maintenance costs, potentially offsetting the savings achieved through energy efficiency. Despite these limitations, the potential for mDWCs to revolutionize multicomponent separations remains compelling. Ongoing research aims to address these challenges through innovations in materials science, computational modeling, and process control. For example, the development of advanced simulation tools, such as Aspen Plus and gPROMS, has facilitated the design and optimization of mDWCs, providing insights into their behavior under various operating conditions (Delgado-Delgado et al., 2012).

The future of mDWCs lies in their integration with emerging technologies, such as artificial intelligence and machine learning. These tools can enhance the design and operation of mDWCs by identifying optimal configurations, predicting performance under varying conditions, and automating control systems. Additionally, advances in sensor technology, including real-time composition analyzers and pressure sensors, can improve operational monitoring and control. Another promising avenue is the hybridization of mDWCs with other separation technologies, such as membranes or adsorption units. These hybrid systems could leverage the strengths of each technology to achieve even greater efficiency and sustainability (Castillo-Landero et al., 2018). For instance, coupling an mDWC with a membrane unit could facilitate the separation of azeotropic mixtures, overcoming one of the key limitations of distillation alone (Castillo-Landero et al., 2018).

Topology and design of a DWC with multiple walls

The industrial implementation of mDWCs has not yet reached the maturity of single-wall DWCs, primarily due to the increased complexity in their design, construction, and operation. While the thermodynamic and economic advantages of mDWCs are well-documented in academic studies, the practical realization of such systems faces several significant barriers. One of the key challenges is the lack of extensive industrial experience with mDWCs. Unlike single-wall DWCs, which are now considered a state-of-the-art technology with widespread adoption in sectors like petrochemicals and fine chemicals, mDWCs require precise engineering and advanced modeling techniques to ensure stable operation. The increased degrees of freedom in mDWCs, often exceeding 20 for four-product configurations, make the design process substantially more intricate. This complexity increases the likelihood of operational instability, which can deter industries from adopting the technology without sufficient proof of feasibility and reliability at a large scale (Lestak & Collins, 1997).

Another barrier is the mechanical and hydraulic challenges posed by the physical implementation of multiple dividing walls. These walls must be precisely aligned and constructed to withstand the dynamic forces within the column. Any misalignment or deformation can compromise the isolation of vapor and liquid flows, leading to suboptimal separation and reduced efficiency. Furthermore, the height of mDWCs, often greater than their single-wall counterparts, poses additional challenges for manufacturing, transportation, and installation, especially in facilities with height restrictions (Rong et al., 2004).

Despite these limitations, the scientific community has shown intense interest in mDWCs due to their outstanding performance in PI studies. Simulations and pilot-scale experiments have demonstrated that mDWCs can achieve energy savings of up to 55% compared to conventional distillation sequences, with significant reductions in equipment and utility costs. This performance advantage has spurred extensive research into overcoming the practical challenges associated with their implementation. Studies are exploring innovative design simplifications, such as reduced-wall configurations, advanced control strategies, and hybrid systems combining mDWCs with other separation technologies (Yildirim et al., 2011). The combination of significant energy savings and reduced environmental impact aligns with the goals of sustainability and green chemistry, further fueling interest in mDWCs. Researchers are particularly focused on their potential applications in industries requiring complex multicomponent separations, such as biofuel production, pharmaceutical synthesis, and the recovery of high-value chemicals from waste streams (Rong, 2011). These applications underscore the importance of continued research to bridge the gap between academic insights and industrial adoption.

The physical construction of an mDWC relies on incorporating multiple dividing walls within a single column shell. These walls create isolated zones, each dedicated to a specific separation task, allowing for simultaneous processing of multiple feed streams and production of high-purity products. The design includes essential elements such as structured packing, weirs, and distributors to manage vapor and liquid flows, as well as standard components like reboilers and condensers. The column shell is typically taller than traditional distillation setups, and its internals are designed to optimize mass and heat transfer while minimizing pressure drops and hydraulic instabilities (Halvorsen & Skogestad, 2003).

The configuration of the dividing walls in an mDWC can vary depending on the desired separation and the feed composition. Common configurations include:

- 2–3–4 Configuration: Features three dividing walls, separating the column into a prefractionator (handling two streams), a middle column (handling three streams), and a main column (handling four streams). This configuration is ideal for separating a four-component mixture into pure products (Halvorsen & Skogestad, 2003)
- 2–2–4-a Configuration: Simplifies the design by eliminating one dividing wall between the middle and main columns. This configuration reduces the number of degrees of freedom, making it easier to control while maintaining high separation efficiency (Dejanović et al., 2011)
- Asymmetric wall arrangements: In some cases, the walls may be positioned asymmetrically to accommodate mixtures with uneven component concentrations or azeotropic behavior. This flexibility allows the column to adapt to challenging feedstocks (Ramapriya et al., 2016)

Each configuration has unique implications for internal flow dynamics, including the distribution of vapor and liquid streams, which must be carefully managed to avoid inefficiencies or operational instability.

The presence of multiple dividing walls introduces significant implications for the distribution and behavior of vapor and liquid flows within the column. Unlike traditional distillation systems, where vapor and liquid freely interact across the column's cross-section, mDWCs enforce strict separation of these streams in different sections. This controlled flow distribution enhances separation efficiency but also adds complexity to the design and operation. Vapor flows are particularly sensitive to pressure drop variations across the dividing walls. Each wall acts as a physical barrier, creating independent pathways for vapor to ascend through the column. The distribution of vapor between these pathways is influenced by the pressure gradients and the flow resistance of the internals, such as trays or structured packing. Mismanagement of vapor flow can lead to flooding in some sections while starving others, reducing the column's overall performance. Advanced computational tools, including CFD, are often used to model and predict these interactions, allowing for design optimization (Preißinger et al., 2019).

Liquid flows are similarly affected by the presence of dividing walls. The flow of liquid between sections is governed by weirs and distributors, which must be carefully designed to ensure even distribution and prevent liquid maldistribution. Uneven liquid flow can cause localized inefficiencies, such as poor contact between vapor and liquid phases, leading to reduced mass transfer rates. Experimental studies have shown that precise control of liquid distribution is essential for achieving the desired separation performance in mDWCs (Wolff & Skogestad, 1995). The thermal gradients introduced by heat transfer across the dividing walls add another layer of complexity. These gradients can influence vapor and liquid compositions, requiring careful design and operation to maintain stability. Despite these challenges, the controlled isolation of vapor and liquid streams allows mDWCs to achieve high separation efficiencies and reduced energy consumption, making them a promising technology for complex multicomponent separations (Lukač et al., 2019).

The control of mDWCs presents unique challenges due to their increased degrees of freedom and the intricate interactions between internal flows. Effective control is critical to maintaining the column's stability and achieving the desired product specifications, particularly in industrial settings where feed compositions and operating conditions may vary. The primary difficulty lies in the lack of direct access to the internal sections of the column. Unlike conventional systems, where control loops can be easily implemented for each column, mDWCs require advanced strategies to monitor and adjust internal flows. Temperature sensors, flow meters, and other instrumentation are often placed at strategic points within the column, but these devices may not capture the full complexity of the internal dynamics. As a result, model-based control strategies, such as MPC, are commonly employed to predict and optimize the column's behavior in real-time (Harvianto et al., 2019). Another challenge is the strong coupling between control variables. For example, adjusting the vapor split in one section can impact the temperature and composition in other sections, creating a ripple effect throughout the column. This interdependence requires sophisticated control algorithms capable of handling multivariable interactions. Additionally, the narrow operating windows of mDWCs mean that even small deviations from optimal conditions can lead to significant performance losses, underscoring the importance of robust and precise control systems (Ramapriya et al., 2016).

The capital and utility savings associated with mDWCs are among their most compelling advantages. By integrating multiple separation stages into a single column shell, mDWCs eliminate the need for additional equipment, such as secondary distillation columns, heat exchangers, and pumps. This consolidation reduces installation costs, minimizes the equipment footprint, and simplifies plant layouts, making mDWCs an attractive option for facilities with limited space. In terms of utility consumption, mDWCs offer remarkable energy efficiency. The reduction in mixing and remixing of streams decreases the reboiler and condenser duties, translating to lower energy costs. Studies have reported utility savings of up to 55% compared to conventional distillation sequences, highlighting the potential for significant cost reductions and environmental benefits. These advantages make mDWCs a promising technology for industries seeking to improve sustainability and reduce operational expenses (Lestak & Collins, 1997).

DWC with multiple walls applied to the separation of complex mixtures

Case study: Separation of hydrocarbon mixtures

Distillation is the most widely used separation process in the chemical industry, valued for its versatility and maturity. However, its high energy consumption and associated greenhouse gas (GHG) emissions highlight the need for innovative solutions within a sustainable process framework. DWCs, as an intensified distillation technology, offer significant improvements by consolidating multiple separation steps into a single column shell. These columns have demonstrated reductions of up to 25% in investment costs, 35% in operating costs, and 40% in space requirements compared to conventional setups (Schultz et al., 2002).

The utility of DWCs extends beyond ternary mixtures, with researchers proposing designs for multicomponent separations. For instance, Rong (2011) introduced a systematic approach for synthesizing DWCs for four-component systems, while Halvorsen and Skogestad (2003) provided analytical solutions to minimize energy use in N-component separations. These advancements underscore the potential of DWCs to deliver substantial energy savings and lower total annualized costs compared to traditional configurations. Studies, such as those by Dejanovic et al. (2010), have shown that DWCs with multiple partition walls can reduce TAC by up to 48%. Despite these advancements, challenges remain. One significant hurdle is the control of vapor splits, which are influenced by mechanical resistances and pressure drops within the column, limiting their adjustability during operation (Harvianto et al., 2019). Researchers like (Ramapriya et al., 2016) have proposed alternative designs, incorporating liquid-only transfers to enhance controllability and dynamic performance. These approaches aim to address inherent safety concerns and operational stability, ensuring that DWCs operate efficiently under varying conditions. This section aims to analyze and evaluate the design and performance of advanced DWCs, particularly those with multiple walls, for quaternary mixtures. By considering sustainability metrics such as energy efficiency, GHG emissions, dynamic behavior, and inherent safety, this analysis seeks to provide a comprehensive understanding of their feasibility and potential for industrial implementation.

Design and optimization of quaternary DWC

The methodology used for the design and optimization of quaternary DWCs revolves around reproducing and simulating separation schemes using Aspen Plus (Fig. 8.3). The feed analyzed corresponds to a light naphtha fraction, with components presenting standard volatilities that include two isomers (2-methylbutane and pentane). This mixture introduces challenges in separation due to the close volatilities of the isomers, making it an ideal test case for

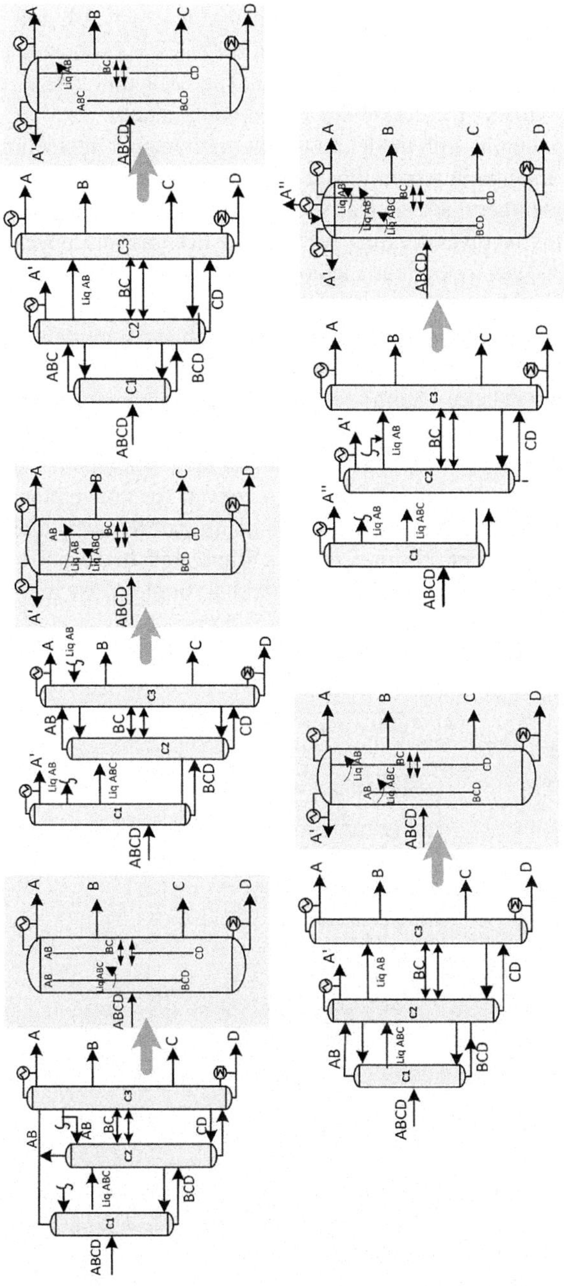

FIGURE 8.3 Quaternary dividing wall column with two walls used as a case of study.

advanced distillation configurations. The feed stream is defined as a saturated liquid at 298 K and includes four hydrocarbons: butane, 2-methylbutane, pentane, and hexane. Their molar fractions are 0.1, 0.4, 0.4, and 0.1, respectively, with purity targets for the products set at 99% for butane and hexane, and 98% for 2-methylbutane and pentane. The total flow rate is 100 kmol/h, with recovery goals of at least 98% for all components.

The design begins with modeling the thermodynamic interactions using the Chao-Seader method in Aspen Plus. Since the simulator lacks a direct module for DWCs, the thermodynamically equivalent Petlyuk column structure is employed. This involves creating preliminary designs for conventional distillation columns using shortcut methods (Underwood, Fenske, and Gilliland equations) to determine the initial parameters such as reflux ratios, the number of stages, and feed locations. Once the initial parameters are defined, the RADFRAC module in Aspen Plus is used to solve the full MESH equations (mass balance, energy balance, thermodynamic model, and composition constraints) for the conventional distillation columns. Purity and recovery targets are adjusted to ensure that the base design satisfies operational requirements. Thermal couplings are then introduced to convert the conventional configurations into DWCs, effectively reducing the number of reboilers and condensers.

The parametric optimization process illustrated in Fig. 8.4 serves as a systematic method to determine the optimal liquid (FL) and vapor (FV)

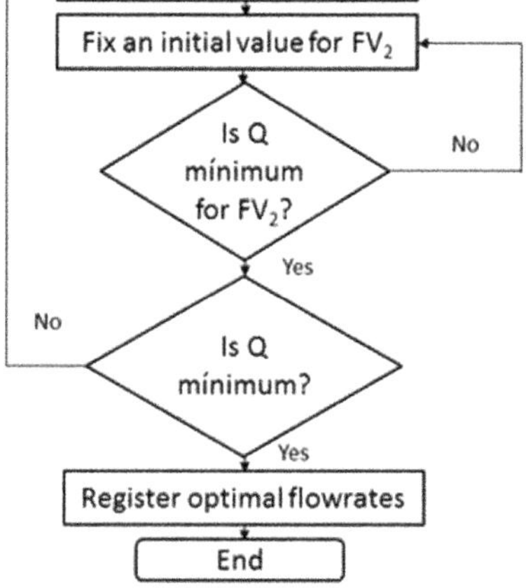

FIGURE 8.4 Parametric optimization procedure.

interconnection flows in a DWC with the objective of minimizing total energy consumption (Q). This process begins by fixing the design parameters of the column, which are derived from the preliminary designs obtained through short-cut methods such as Underwood, Fenske, and Gilliland. These parameters include the number of stages, feed locations, and interconnection points. The design aims to ensure that product purity targets of 99 mol% for components A (butane) and D (hexane) and 98 mol% for components B (2-methylbutane) and C (pentane) are achieved, with recovery rates of at least 98% for all components.

Once the design parameters are fixed, the optimization process initiates by assigning initial values to the liquid (FL1) and vapor (FV2) flows in the interconnections. These initial values are used as starting points for evaluating the energy requirements of the column. The process then involves iterative adjustments to the interconnection flows to identify local and global energy minima. First, the vapor flow FV is varied systematically to identify a local energy minimum while keeping the liquid flow FL1 constant. This step ensures that the vapor flow is optimized for a specific liquid flow value. After identifying the minimum for FV2, the process moves to optimize the liquid flow FL. The liquid flow is adjusted, and the vapor flow is reevaluated to identify new energy minima. This iterative cycle of adjusting liquid and vapor flows continues until the global energy minimum for the system is identified.

Throughout this process, the energy consumption at each combination of FL1 and FV2 is calculated and compared. The goal is to identify the specific combination of liquid and vapor interconnection flows that minimizes the reboiler duty and overall energy requirements of the column. Once the optimal values of FL1 and FV2 are determined, they are recorded as the ideal operating parameters for the column. The methodology provides a practical approach to optimizing complex distillation systems such as DWCs, which involve multiple degrees of freedom due to thermal couplings and interconnections. While the parametric optimization method may not be as robust as other advanced optimization techniques, it has been shown to be effective in identifying optimal operating zones for complex column configurations. The systematic nature of this method allows for the exploration of a wide range of operational conditions, ensuring that the energy requirements are minimized while maintaining the desired product purity and recovery levels.

This approach is particularly beneficial for DWCs, where the inclusion of thermal couplings and interconnections introduces challenges in balancing liquid and vapor flows. The optimization process accounts for these complexities by iteratively refining the operational parameters to achieve energy efficiency. Moreover, the method is adaptable to various feed compositions and column configurations, making it a versatile tool for the design and optimization of advanced distillation systems. By minimizing the energy consumption, the parametric optimization not only reduces operational costs but also contributes to sustainability by lowering GHG emissions associated with energy use. This makes it a valuable methodology for the design and operation of DWCs in a framework of sustainable chemical engineering.

Process assessment

The evaluation of quaternary DWCs was carried out using critical sustainable indicators, namely GHG emissions, inherent safety, and dynamic behavior under control conditions. These indicators provide a comprehensive understanding of the performance of the columns, not only in terms of energy efficiency but also regarding their environmental impact and operational safety.

Greenhouse gas emissions were calculated following the methodology proposed by Gadalla et al. (2005), which links CO_2 emissions directly to the energy demand of the reboiler. The model assumes complete combustion of fuel, avoiding CO emissions, and quantifies CO_2 based on the carbon content and net heating value (NHV) of the fuel used (Eq. 8.1). Methane, with an NHV of 49,675 kJ/kg, was used as the reference fuel in this study. This indicator is critical because reducing the energy consumption of a column directly lowers its carbon footprint, aligning the process with global sustainability goals. CO_2 emissions are computed as follows:

$$[CO_2] = (Q_{FUEL}/NHV)\left((C\%)/100\right)\alpha \tag{8.1}$$

where Q_{fuel} (kW) is the amount of fuel burned, $\alpha = 3.67$ is the ratio of molar masses of CO_2 and C, and NHV (kJ/kg) represents the NHV of a fuel with carbon content of C%.

The inherent safety of each configuration was quantified through the Individual Risk (IR) metric, a probabilistic measure of the risk to a person located within the process plant. IR calculations consider accident frequency, probability of damage, and the severity of potential incidents such as boiling liquid expanding vapor explosions, vapor cloud explosions, and toxic releases. The risk analysis was informed by quantitative risk assessment, which identifies possible accidents, their causative variables (e.g., thermal radiation, overpressure, concentration), and their effects at a fixed distance of 50 m. Probit models were employed to evaluate the probability of injury or death based on exposure to these variables, providing a robust framework for assessing safety performance (Crowl & Louvar, 2001). IR index is calculated as shows Eq. 8.2.

$$IR = \Sigma\, f_i P_{x,y} \tag{8.2}$$

The importance of this metric lies in its ability to ensure that advanced distillation systems like DWCs do not compromise operational safety while pursuing energy and cost efficiency.

Dynamic behavior was assessed through a closed-loop control analysis in Aspen Dynamics to evaluate the response of each scheme under two scenarios: a set-point change in product composition and a disturbance in the molar feed flow. Using proportional-integral controllers, the system's ability to maintain product composition under such conditions was measured by the Integral of Absolute Error (IAE) criterion. The manipulated variables for control included reflux flow rate (L), vapor boil-up rate (V), and side stream flow rate (S), selected for their direct influence on composition control. The control strategy ensured stable operation, avoiding composition drifts that could affect product purity and overall performance. This dynamic evaluation is vital as it verifies the operational robustness of the column designs, particularly under thermally coupled configurations, and ensures minimal waste and consistent output quality during normal operation and process disturbances.

The use of GHG emissions, inherent safety, and dynamic performance as evaluation metrics provides a holistic approach to assessing the sustainability of DWCs. These indicators collectively address the environmental, safety, and operational dimensions of the process. By ensuring that the columns meet stringent requirements in these areas, the evaluation framework facilitates the design of DWCs that are not only energy-efficient but also environmentally friendly, inherently safe, and dynamically robust. Such a comprehensive assessment is crucial for encouraging industrial adoption of advanced distillation systems in a sustainable framework.

Results

All the designs depicted in Fig. 8.3a–f were created using Aspen Plus. Prior to conducting any evaluations, it was essential to optimize each design to ensure a fair comparison among the schemes, with all configurations operating at their respective minimum energy requirements (Table 8.1).

On the other hand, the design tables for all schemes are shown below (Tables 8.2–8.6).

Table 8.1 highlights the differences in inherent safety across all the evaluated schemes, showing that designs incorporating liquid splits are inherently

TABLE 8.1 CO_2 emissions and inherent safety.

Scheme	Reboiler duty (kJ/s)	CO_2 emissions (Ton/h)	IR (Probability/y)
A	6562.31	32.775	2.7431E-04
B	6516.92	32.548	2.7351E-04
C	6514.12	32.534	2.7502E-04
D	6504.98	32.488	2.6747E-04
E	6548.27	32.705	2.6029E-04
F	6510.39	32.515	2.7605E-04

TABLE 8.2 Design parameters for scheme A.

	Pre-frac II	Pre-frac I	Main column
Number of stages	35	42	118
Feed stage	1, 6, 35	1, 7, 16, 32, 42	11, 55, 55, 73
Top temperature (K)	349.52	351.54	320.45
Bottom temperature (K)	355.22	365.87	405.06
Top pressure (kPa)	367.11	471.23	436.65
Bottom pressure (kPa)	371.32	540.45	574.7
Reflux rate (kmol//h)	N/A	N/A	1265.12
Reboiler duty (kJ)	N/A	N/A	6562.31
Product stage	1, 6, 35	1, 13, 31, 42	1, 10, 13, 13, 20, 72, 101, 118
ABC liquid flow (kmol//h)	12.006	AB vapor/liquid flow (kmol/h)	48.9867/13
BCD vapor/liquid flow (kmol//h)	44/93.4133	BC vapor/liquid flow (kmol//h)	22/73
CD vapor/liquid flow (kmol/h)	80/118.225		

safer than those relying solely on conventional thermal couplings. While the reboiler heat duty plays a significant role in determining inherent safety, it is only one of many factors considered in the calculation. Operative pressure, for instance, also influences the IR value. However, since all schemes were designed to operate under the same pressure, this variable does not contribute to differences in safety among the configurations. The primary variations in IR values, as detailed in Tables 8.2–8.6, stem from the distribution of liquid and vapor flows in conventional thermal couplings and the novel liquid splits. These differences directly impact the amount of material present within the column, which is a key factor in determining inherent safety. The explanation is straightforward: the quantity of material in each section of the column, influenced by the liquid and vapor flows, directly affects the IR values. Consequently, the design choice between conventional thermal couplings and liquid splits significantly influences the overall safety of the column, with the latter offering a safer alternative.

When evaluating the dynamic behavior of all schemes, it was observed that the liquid-split designs demonstrated a superior ability to handle setpoint changes and disturbances compared to the conventional scheme, which relied solely on thermal couplings. Among all configurations, Scheme E exhibited the best dynamic performance under both scenarios.

TABLE 8.3 Design parameters for scheme B.

	Pre-frac II	Pre-frac I	Main column
Number of stages	10	42	118
Feed stage	9, 10	1, 7, 10, 32, 42	4, 5, 11, 72
Top temperature (K)	314.09	347.2	320.45
Bottom temperature (K)	348.78	370.33	404.45
Top pressure (kPa)	367.11	471.23	437.21
Bottom pressure (kPa)	436.22	540.54	574.35
Reflux rate (kmol//h)	56.09	N/A	1213.95
Reboiler duty (kJ)	N/A	N/A	6516.92
Product stage	1, 2, 3, 10	1, 13, 42, 31	1, 3, 10, 20, 72, 101, 118
AB liquid flow (kmol/h)	1.3	ABC liquid flow (kmol/h)	2.89
BCD vapor/liquid flow (kmol/h)	27/126.159	BC vapor/liquid flow (kmol/h)	31/2
AB vapor/liquid flow (kmol/h)	74.9252/55	CD vapor/liquid flow (kmol/h)	85/118.225

As shown in Fig. 8.3, the primary physical distinction of Scheme E lies in its increased substitution of liquid splits compared to the other novel schemes. Additionally, the vapor flow associated with each LV thermal coupling in Scheme E is the highest among all evaluated configurations, including those with conventional thermal couplings. Another notable feature of Scheme E is its ability to purify the lightest component in three separate streams at the top of the column, further contributing to its enhanced performance. Table 8.7 shows a summary regarding dynamic evaluation.

Scheme E demonstrated the best capacity to handle disturbances among all configurations. As shown in Fig. 8.3, this superior performance is primarily attributed to its design, which incorporates a greater number of liquid splits compared to the other novel schemes. Additionally, Scheme E exhibits the highest vapor flow in each LV thermal coupling compared to all conventional thermal couplings. A key operational feature of Scheme E is its ability to purify the lightest component in three separate streams at the top of the column, which contributes to its enhanced dynamic and separation performance.

The reported results highlight the significant implications of designing complex systems with liquid splits rather than traditional thermal couplings.

TABLE 8.4 Design parameters for scheme C.

	Pre-frac II	Pre-frac I	Main column
Number of stages	10	42	118
Feed stage	1, 9, 16	3, 12, 22, 30	3, 15, 27
Top temperature (K)	339.53	323.353	320.46
Bottom temperature (K)	34.04	364.285	404.938
Top pressure (kPa)	669.76	640.37	640.37
Bottom pressure (kPa)	675.84	654.56	655.57
Reflux rate (kmol/h)	N/A	217.63	77.54
Reboiler duty (kJ)	N/A	N/A	6514.12
Product stage	1, 16	1, 3, 7, 12, 12, 30	1, 7, 15, 23, 27, 41
ABC vapor/liquid flow (kmol/h)	31.439/25	BC vapor/liquid flow (kmol/h)	31/17
BCD vapor/liquid flow (kmol/h)	25/118.561	CD vapor/liquid flow (kmol/h)	85/176.318
AB liquid flow (kmol/h)	20.021		

Notably, designs with a higher number of liquid splits were found to be thermodynamically equivalent to those with fewer splits or even conventional thermal couplings, demonstrating no additional energy demand for this topology. However, the advantages of liquid splits were most evident in terms of dynamic behavior and inherent safety. Scheme E, for instance, achieved the lowest IR value, which is consistent with its dynamic performance. A well-controlled process minimizes waste and reprocessing needs, ensures consistent product quality, and maintains stable operating conditions, thereby avoiding risky scenarios with potentially severe consequences.

Thus, incorporating liquid splits into the design not only improves the dynamic behavior and inherent safety of the process but does so without increasing energy consumption or environmental impact. This makes liquid splits a highly advantageous feature for complex distillation schemes.

Case study: Separation of biobutanol from fermentation

Biobutanol has emerged as a promising alternative to bioethanol in the quest for sustainable biofuels. Its higher energy density, reduced water absorption,

TABLE 8.5 Design parameters for scheme D.

	Pre-frac II	Pre-frac I	Main column
Number of stages	10	42	118
Feed stage	1, 6, 10	4, 7, 11, 32, 42, 42	4, 11, 73
Top temperature (K)	334.898	323.357	320.459
Bottom temperature (K)	347.898	361.031	405.689
Top pressure (kPa)	669.76	640.37	640.37
Bottom pressure (kPa)	675.84	655.57	654.56
Reflux rate (kmol/h)	N/A	241.9	69.65
Reboiler duty (kJ)	N/A	N/A	6504.98
Product stage	1, 4, 10	1, 9, 12, 13, 31, 42	1, 10, 20, 72, 101, 118
AB vapor/liquid flow (kmol/h)	14.4551/10	AB liquid flow (kmol/h)	3.001
ABC liquid flow (kmol/h)	3.002	BC vapor/liquid flow (kmol/h)	31/5
BCD vapor/liquid flow (kmol/h)	10/102.545	CD vapor/liquid flow (kmol/h)	95/216.533

and compatibility with existing fuel distribution infrastructure make it a competitive option for replacing or supplementing gasoline. However, biobutanol production faces challenges, including lower yields compared to bioethanol—0.11 kg of biobutanol per kg of corn versus 0.30 kg of bioethanol—highlighting the need for innovations to improve overall process efficiency (Karimi et al., 2015). The production process for biobutanol, like most alcohols obtained via fermentation, can be divided into three main stages: biomass pretreatment, fermentation, and product removal and purification (Amiri & Karimi, 2015). While the pretreatment step aims to maximize fermentable sugar yields through various methods, including physical, chemical, or biological processes, the fermentation step relies on Clostridium bacteria. Advances in metabolic engineering have significantly enhanced biobutanol production during this phase (Jafari et al., 2016).

The product removal and purification stage presents significant challenges due to the thermodynamic complexities of the fermentation broth. The ABE (acetone-butanol-ethanol) mixture includes azeotropes—homogeneous between ethanol and water, and heterogeneous between butanol and water—which make separation by distillation highly energy-intensive. To address this, researchers

TABLE 8.6 Design parameters for scheme E.

	Pre-frac II	Pre-frac I	Main column
Number of stages	10	42	118
Feed stage	6, 10	4, 7, 42, 32	4, 4, 11, 73
Top temperature (K)	273.639	310.274	320.46
Bottom temperature (K)	308.561	351.11	404.894
Top pressure (kPa)	669.76	640.37	640.37
Bottom pressure (kPa)	678.88	654.56	654.56
Reflux rate (kmol/h)	46.48	133.86	70.99
Reboiler duty (kJ)	N/A	N/A	6548.27
Product stage	1, 2, 7, 10	1, 9, 13, 42, 31	1, 10, 20, 72, 101, 118
AB liquid flow (kmol/h)	1.02	AB liquid flow (kmol/h)	5.03
ABC liquid flow (kmol/h)	5.01	BC vapor/liquid flow (kmol/h)	10/4
BCD vapor/liquid flow (kmol/h)	100/192.171	CD vapor/liquid flow (kmol/h)	168/264.83

have proposed a range of separation alternatives, including liquid-liquid extraction, gas stripping, and distillation hybrids. Among these, liquid-liquid extraction combined with distillation has proven to be particularly efficient, as it mitigates butanol inhibition during fermentation and reduces energy requirements (Contreras-Vargas et al., 2019). DWCs offer a promising solution for optimizing biobutanol separation. DWCs have already demonstrated significant savings in bioethanol purification, with studies reporting up to 60% energy savings and 20% reductions in TAC compared to conventional setups. For biobutanol, their application remains underexplored (Ignat & Kiss, 2013), but studies suggest that DWCs could achieve similar benefits.

This section aims to present and analyze DWC-based alternatives for biobutanol separation, exploring both conventional and nonconventional designs. By employing a multiobjective framework that incorporates economic, environmental, and controllability metrics, the goal is to identify optimal configurations for the sustainable and efficient recovery of biobutanol from fermentation broths.

TABLE 8.7 IAE values for set point change.

	Component A			Component B			Component C			Component D			Component A′			Component A″		
	K	Ti	IAE	K	Ti	IAE	K	Ti	IAE	K	Ti	IAE	K	Ti	IAE	K	Ti	IAE
	(%mol/ %mol)	(min)	(–)	(%mol/ %mol)	(min)	(–)	(%mol/ %mol)	(min)	(–)	(%mol/ %mol)	(min)	(–)	(%mol/ %mol)	(min)	(–)	(%mol/ %mol)	(min)	(–)
Scheme A	250	1	7.90E-05	250	40	3.43E-03	250	50	4.10E-03	250	1	8.05E-05	–	–	–	–	–	–
Scheme B	250	1	7.90E-05	250	1	1.40E-03	250	10	3.22E-03	250	1	8.00E-05	–	–	–	250	10	8.29E-04
Scheme C	250	1	8.06E-05	250	1	1.28E-03	250	40	3.36E-03	240	1	7.96E-05	250	1	1.01E-04	–	–	–
Scheme D	250	1	8.09E-05	250	1	8.96E-04	250	30	3.34E-03	250	1	8.11E-05	250	1	8.03E-05	–	–	–
Scheme E	250	1	7.90E-05	250	1	1.19E-03	250	30	3.27E-03	250	1	8.12E-05	250	1	2.44E-04	250	1	8.40E-05
Conventional	250	5	4.11E-04	250	150	1.23E-02	250	10	3.39E-03	250	1	1.16E-04						

Hybrid flowsheets synthesis procedure

The synthesis procedure is a critical step in generating a comprehensive search space of potential configurations for multicomponent separations. This approach prevents reliance on preexisting configurations that might not fully address the specific case under consideration and ensures no potentially optimal solutions are overlooked. Sequential synthesis procedures, beginning with simple column arrangements, have proven effective for developing complex multicomponent configurations. Simple column sequences, while straightforward, offer various arrangements based on the separation order of components in multicomponent mixtures. In the case of hybrid liquid-liquid extraction-assisted distillation flowsheets, the extracted phase is introduced to the distillation section. For ABE separation, this feed stream typically consists of a four-component mixture containing solvent, acetone, butanol, and ethanol. Five possible simple column sequences can be generated for this system.

For intensified distillation alternatives such as DWCs, a distinction is made between conventional and nonconventional configurations. Conventional DWCs, as defined by Kaibel (1987), consist of a single-shell column equipped with a vertical partition, a condenser, and a reboiler, capable of generating three or four product streams. In this section, the case of three streams is considered. The design of nonconventional DWCs, however, requires a more systematic synthesis methodology. Nonconventional DWCs are characterized by additional features such as multiple reboilers, condensers, intermediate reboilers, or multiple walls. Rong (2011) proposed a four-step procedure to systematically generate a subspace of nonconventional DWCs starting from simple column configurations. The synthesis process involves the following steps: First, a simple column configuration is selected from the complete set of possibilities. Second, thermally coupled configurations are generated by replacing condensers and/or reboilers associated with nonproduct streams with bidirectional thermal couplings of vapor and liquid streams. Finally, these equivalent structures are transformed into multicomponent DWCs by integrating single-column sections into their thermally linked columns using dividing walls. This methodology not only generates a complete set of alternatives but also offers the advantage of establishing a direct correspondence between simple column configurations and their derived DWCs. Once the optimal simple column configuration is identified, only the DWCs derived from that sequence need to be evaluated, streamlining the design process and focusing efforts on the most promising configurations.

Design and optimization procedure

The design of hybrid liquid-liquid extraction-assisted distillation flowsheets was guided by minimizing an objective function incorporating three critical indices: TAC, Eco-Indicator 99 (EI99), and the condition number (CN). This multiobjective approach ensures economic feasibility, minimizes environmental impact, and enhances the controllability of the process, thereby guaranteeing smooth plant operation even under disturbances. All model used for objective function calculation have been approached in previous chapters.

The design of the proposed alternatives was conducted by minimizing a multiobjective function (Eq. 8.3):

$$\mathrm{Min}(\mathrm{TAC}, \mathrm{EI99}, \mathrm{CN}) = \mathrm{f}\left(\mathrm{N_e}, \mathrm{N}, \mathrm{N_f}, \mathrm{RR}, \mathrm{D}, \mathrm{F_L}, \mathrm{F_V}, \phi, \mathrm{S}\right) \tag{8.3}$$

Here, N_e represents the number of stages in the extractor, N the number of stages in the distillation columns, N_f the feed stage, RR the reflux ratio, D the distillate flow rate, F_L and F_V the liquid and vapor interconnection flows, ϕ the column diameter, and S the solvent flow rate. The function is subject to the constraint $y \geq x$, where y and x are the obtained and required purity vectors, respectively. Given the complexity of the optimization problem, a differential evolution with tabu list (DETL) algorithm was employed. This stochastic optimization technique has been demonstrated to effectively handle multiobjective problems in various engineering contexts (Rangaiah & Pandu, 2010). The DETL algorithm operates through four main steps: initialization, mutation, crossover, and selection. To implement the optimization process, Aspen Plus was integrated with Microsoft Excel using COM-based dynamic data exchange. Design variables were transferred to Excel, where they were mapped to the corresponding process variables in Aspen Plus. After completing the simulation, Aspen Plus returned results to Excel. A macro coded in Visual Basic within Excel then analyzed the objective function values and generated new decision variable values based on the DETL method. The optimization process was configured with the following control parameters: 200 individuals, 500 generations, a tabu list containing 50% of the total population, a taboo radius of 10^{-6}, a crossover probability of 0.8, and a mutation factor of 0.6. These parameters were fine-tuned through preliminary simulations to ensure all purity constraints were met for each component. The final solution was selected using a Pareto-based approach, ensuring an optimal balance between the economic, environmental, and controllability objectives. This methodology provided a robust framework for optimizing the complex design space of hybrid extraction-distillation systems.

To compare the different hybrid flowsheets, a feed stream of 1.64 kmol/h was analyzed, containing 8.1% acetone, 11.3% butanol, 0.4% ethanol, and 80.2% water (molar basis) at 35°C and 1 atm. The feed composition is based on Sánchez-Ramírez et al. (2015). All proposed configurations were simulated using Aspen Plus, employing the NRTL-Hayden O'Connell equation of state with Henry's law as the thermodynamic model. Hexyl-acetate was selected as the mass separating agent. The minimum purity requirements were established on a mass basis, set at 99.5% for acetone and biobutanol, and 99.0% for ethanol. Column operating pressures were optimized considering the availability of cooling water at 20°C, with low, medium, and high-pressure steam utilized as auxiliary fluids for the reboilers based on the temperature of the bottom streams.

The hybrid liquid-liquid extraction configurations assisted by nonconventional DWCs, derived from the HLL-ISC configuration, are presented in Fig. 8.5. These configurations can be grouped into four distinct categories:

These categories capture the diversity of hybrid flowsheet designs and highlight the versatility of nonconventional DWCs in optimizing separation processes.

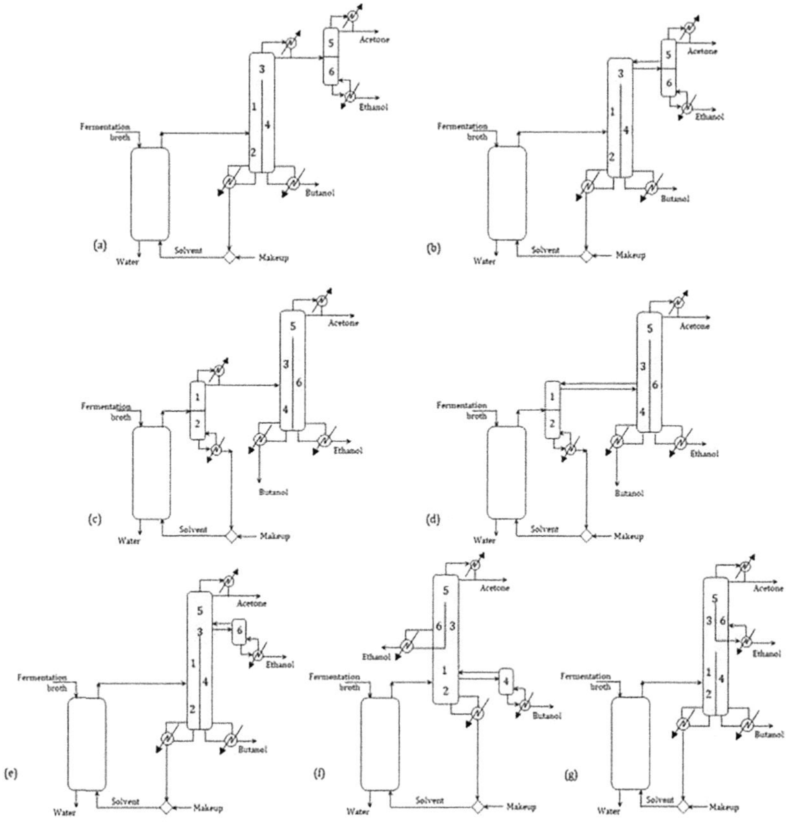

FIGURE 8.5 Liquid-liquid extraction assisted nonconventional DWC configurations. Configuration consisting of a liquid-liquid extractor, a DWC, and a simple column, as shown in Fig. 8.5a and c. Configurations that include a liquid-liquid extractor and a DWC thermally coupled with a simple column, depicted in Fig. 8.5b and d. Configurations comprising a liquid-liquid extractor and a DWC thermally coupled with a single rectifying section, illustrated in Fig. 8.5e and f. Configurations featuring a liquid-liquid extractor and a single DWC with multiple partitions and an intermediate reboiler, as represented in Fig. 8.5g.

Among the liquid-liquid extraction-assisted nonconventional DWCs, the configuration in Fig. 8.5g achieved the lowest TAC and EI99 values, although it is penalized by its controllability. This result aligns with its intensified design, where the solvent and the ABE mixture are separated within the same column, maximizing economic performance. However, when compared to the best-performing configuration from the liquid-liquid extraction-assisted simple column distillation alternatives, shown in Fig. 8.5e, the latter demonstrates superior overall performance. Specifically, the configuration in Fig. 8.5e reduces the TAC and EI99 by 22% and 18%, respectively, while also achieving a better controllability index. From a structural perspective, configurations 8.5e and g differ primarily in column section 6. In the configuration of Fig. 8.5e, section 6 is a single rectifying section, whereas in Fig. 8.5g, it is integrated into the DWC. While the differences in TAC and EI99 between the two configurations are <5%, the higher CN value associated with Fig. 8.5g establishes the configuration in Fig. 8.5e as the optimal choice overall.

Additional configurations also merit attention. Figs. 8.5b and d, where the DWC is thermally coupled with a simple column, exhibit lower objective function values compared to the reference case and differ from the best alternative by <5%. The configurations in Fig. 8.5a and c which consist of a liquid-liquid extractor, a DWC, and a simple column, also demonstrate competitive performance but fall short of the configurations in Fig. 8.5b, d, e. Table 8.8 shows the objective function for all DWC alternatives.

On the other hand, Table 8.9 shows the design parameters for the scheme presented in Fig. 8.5g.

Biobutanol, a promising alternative to bioethanol, is gaining industrial interest despite its production process requiring significant optimization. This study focused on improving the separation of biobutanol, acetone, and ethanol through hybrid liquid-liquid assisted distillation. Innovative configurations combining liquid-liquid extraction with various DWC designs were systematically generated and optimized using a multiobjective function addressing economic (TAC), environmental (EI99), and controllability factors. The best-performing configuration, featuring a DWC with two reboilers and a side rectifying stream, achieved a 22% reduction in TAC, an 18% reduction in EI99, and improved controllability compared to conventional setups. Two other configurations also demonstrated competitive performance, with less than a 5% difference from the best alternative and enhanced controllability. These novel configurations, applied to ABE separation for the first time, present a tangible opportunity to improve the efficiency and sustainability of biobutanol production, boosting its industrial viability.

Case study: Production and separation of lactic acid

Lactic acid (2-hydroxypropionic acid) is a versatile organic compound with significant industrial relevance. It exists in two optically active isomeric forms,

TABLE 8.8 Objective function values for the configurations of Fig. 8.5 (Errico et al., 2016).

Objective function	Fig. 8.5a	Fig. 8.5b	Fig. 8.5c	Fig. 8.5d	Fig. 8.5e	Fig. 8.5f	Fig. 8.5g
TAC (k$/yr)	108.54	105.57	115.5	101.78	100.85	100.59	97.88
EI99 (kpoints/yr)	13.73	12.93	14.34	13.3	12.79	14.74	12.22
CN	1402	1.7	$1.22 \cdot 10^{17}$	3.9	7.3	9888.3	18,994.4

TABLE 8.9 Design parameters for the scheme presented in Fig. 8.5.

		Column sections			
	Extractor	1 + 2	4	5 + 3	6
Number of stages	5	43	43	71	7
Feed location	–	13	–	–	–
Reflux ratio	–		–	0.644	–
Distillate flowrate (kg/h)	–	–	–	7.717	–
Residue flowrate (kg/h)	–	712.11	13.68	–	0.32
Liquid split flowrate (kg/h)	–	43.463	17.38	–	–
Vapor split flowrate (kg/h)	–	–	–	–	–
Extract flowrate (kg/h)	733.873	–	–	–	–
Solvent flowrate (kg/h)	712.147	–	–	–	–
Diameter (m)	0.335	0.288	0.288	0.379	0.3
Pressure (kPa)	101.3	101.3	101.3	101.3	101
Condenser duty (kW)	–	0	0	11.23	0
Reboiler duty (kW)	–	69.92	0.633	0	0.02

L(+) and D(−), and is widely utilized across various sectors. Lactic acid serves as a feedstock in the production of biodegradable plastics like polylactic acid (PLA), lactate esters, and other value-added chemicals. It is also a critical ingredient in the food industry, where it is used as a preservative and flavoring agent, and in pharmaceuticals for the synthesis of biocompatible materials. Additionally, lactic acid finds applications in cosmetics as a humectant for creams and soaps. With an estimated annual global demand between 200,000 and 350,000 metric tons and growth rates of 12%–15%, lactic acid has become indispensable in both established and emerging markets (Abdel-Rahman et al., 2013). Commercially, lactic acid is predominantly produced through fermentation of carbohydrate-rich feedstocks, such as glucose (derived from starch), maltose (from enzymatic starch conversion), sucrose (from syrups or molasses), and lactose (from whey) (Corma Canos et al., 2007). Fermentation offers significant advantages over chemical synthesis, such as high specificity and the ability to produce optically pure isomers, depending on the microorganism used. Notably, the L(+) isomer is of particular importance due to its role in PLA production, a key material in biodegradable plastics (Cavani et al., 2016). However, several challenges hinder the efficiency of the fermentation process. The fermentation broth typically contains dilute concentrations of lactic acid and various impurities, such as unconverted sugars, heavy metals, and proteins. These impurities, combined with lactic acid's high affinity for water and its tendency to oligomerize at elevated temperatures, make the separation and purification processes complex and energy-intensive.

The traditional recovery process for lactic acid involves calcium lactate precipitation, followed by acidification with sulfuric acid to release lactic acid. While this method is well-established, it is far from ideal, accounting for up to 50% of production costs (Cavani et al., 2016). To address these inefficiencies, advanced (Komesu et al., 2015) separation techniques such as solvent extraction, adsorption, short-path evaporation, and membrane processes have been proposed. However, these methods face limitations, including high solvent requirements, low recovery efficiencies, and operational challenges such as membrane fouling (Komesu et al., 2015).

Process intensification (PI) provides a promising framework to overcome these challenges by integrating multiple operations into compact, efficient systems. PI reduces heat and mass transfer resistances, enhances energy efficiency, and minimizes environmental impact. Reactive distillation, a prime example of PI, combines reaction and separation into a single unit, offering advantages such as increased conversion, enhanced selectivity, reduced (Kim et al., 2017) energy consumption, and the ability to overcome azeotropes. For lactic acid purification, reactive distillation can be employed to convert lactic acid into methyl lactate using methanol. This intermediate product is easier to purify via distillation and can subsequently be hydrolyzed back to lactic acid. This approach reduces energy requirements and improves process efficiency, making it a promising solution to lactic acid purification. Another intensified approach involves the use of thermally coupled systems, such as DWCs. These configurations allow for the separation of complex mixtures within a single column shell, achieving significant energy savings and lower operational costs compared to conventional distillation (Van Duc Long & Lee, 2017, p. 1–210). By integrating these advanced separation techniques, PI offers a pathway to sustainable lactic acid production, aligning with the principles of a circular economy where economic, environmental, and societal outcomes are balanced.

The objective of this section is to explore the development of an intensified separation process for lactic acid recovery and purification. This study will focus on the use of reactive distillation to address the challenges posed by the L(+) and D(−) isomers of lactic acid. Furthermore, thermally coupled and intensified separation schemes will be proposed to minimize energy consumption and enhance the sustainability of the lactic acid purification process. By integrating PI principles and sustainability indicators, this work aims to provide innovative and viable industrial solutions for the efficient production of lactic acid (Faraj et al., 2021).

The production and purification of lactic acid, particularly through intensified processes, represent a significant opportunity to enhance efficiency and sustainability in industrial operations. One promising approach involves reactive distillation, which integrates esterification and hydrolysis reactions with methanol, along with managing the oligomerization of lactic acid. This intensified separation strategy offers a viable alternative to conventional distillation, which often struggles with the complex thermodynamic interactions inherent in lactic acid systems.

Preliminary studies by Kim et al. (2017) demonstrated the feasibility of using reactive distillation for lactic acid purification. Following the design guidelines suggested, the process was validated against key requirements: (1) the presence of multiple products; (2) compatibility between reaction and separation conditions, as evidenced by reaction temperatures near 50°C and boiling points of the main reactants, methanol (64°C) and lactic acid (122°C); and (3) operational conditions avoiding the critical region of any component. These criteria confirm the industrial applicability of (Kim et al., 2017) reactive distillation for lactic acid, as described.

The reactions occurring within the columns include esterification to form methyl lactate, as well as oligomerization reactions producing dilactate and trilactate. These reactions can be represented by the following sequences (Fig. 8.6).

From an operational standpoint, controlling the dome temperature is critical, as temperatures exceeding 80°C increase the likelihood of oligomerization. To address this, operational pressure becomes a key variable, maintaining conditions that prevent unwanted reactions while enabling effective separation. The process also assumes that solid catalysts occupy 50% of the tray holdup volume, with a catalyst density of 770 kg/m^3 used for holdup calculations. For nonideal systems with strong vapor-phase interactions, the UNIQUAC-HOC and NRTL-HOC thermodynamic models provide reliable property predictions. In this study, the UNIQUAC-HOC model was employed to ensure consistency with previously published benchmarks. The representative fermentation broth fed into the reactive distillation columns is

$$\underbrace{CH_3CH(OH)COOH}_{Lactic\ Acid} + \underbrace{CH_3OH}_{Methanol} \underset{k_{r,1}}{\overset{k_{f,1}}{\leftrightarrow}} \underbrace{CH_3CH(OH)COOCH_3}_{Metil\ lactate} + \underbrace{H_2O}_{water}$$

$$\underbrace{CH_3CH(OH)COOH}_{Lactic\ Acid} + \underbrace{CH_3CH(OH)COOH}_{Lactic\ Acid} \underset{k_{r,2}}{\overset{k_{f,2}}{\leftrightarrow}} \underbrace{C_6H_{10}O_5}_{Dilactate} + \underbrace{H_2O}_{water}$$

$$\underbrace{C_6H_{10}O_5}_{Dilactate} + \underbrace{CH_3CH(OH)COOH}_{Lactic\ Acid} \underset{k_{r,3}}{\overset{k_{f,3}}{\leftrightarrow}} \underbrace{C_9H_{14}O_7}_{Trilactate} + \underbrace{H_2O}_{water}$$

FIGURE 8.6 Lactic esterification to form methyl lactate.

TABLE 8.10 Characterization feed from fermenter.

Flowrate (kmol/h)	Pressure (atm)	Temperature (°C)	%mol		
			Láctic acid	Water	Succinic acid
50	2	35	8.4	90.5	1.1

characterized in Table 8.10, with a flow rate of 50 kmol/h containing 8.4% lactic acid, 90.5% water, and 1.1% succinic acid at 35°C and 2 atm. This composition facilitating comparisons with earlier studies. The base case for this study involves a conventional distillation scheme, using the fermentation broth as feed. Initially, the broth is preconcentrated in the first column to remove most of the water, avoiding the azeotropic point. The resulting stream is then directed to a second column where esterification with methanol occurs. This step simultaneously produces methyl lactate and separates high-purity succinic acid from the column bottom. The distillate is fed into a third column for hydrolysis, regenerating lactic acid. Finally, the last column purifies methanol and water for recirculation into the process.

This intensified approach, integrating esterification and hydrolysis with reactive distillation, showcases significant potential for improving lactic acid purification by reducing energy consumption and enhancing overall efficiency. By leveraging these advanced configurations, the study explores innovative solutions for industrial-scale lactic acid recovery.

Synthesis of intensified separation alternatives

The synthesis of intensified separation alternatives for lactic acid purification follows a systematic methodology designed to generate highly efficient configurations by integrating thermal couplings and optimizing separation strategies (Errico et al., 2009). This approach transforms conventional distillation systems into advanced intensified alternatives capable of achieving significant energy savings and operational improvements.

The process begins with the identification of a reference configuration, typically based on a conventional distillation scheme, which serves as the foundation for modifications. From this baseline, auxiliary equipment such as reboilers and condensers are progressively replaced by interconnection streams, creating thermally coupled systems. These interconnections enable energy integration and reduce external utility requirements, laying the groundwork for the development of more efficient designs.

The next phase involves the identification of thermodynamically equivalent configurations, where column sections are rearranged to explore alternative separation pathways. This step is particularly challenging for azeotropic mixtures, such as those found in lactic acid purification, as their nonideal behavior adds complexity to the process. By systematically shifting column sections, new configurations are generated, expanding the design space and ensuring that potential optimal solutions are not overlooked.

Finally, the construction of side-stream configurations consolidates the intensified design. By eliminating auxiliary sections, such as side columns, and incorporating side streams, the process achieves simplification without compromising separation efficiency. The placement of these side streams is determined by the position of the removed sections and the final arrangement of separation zones. The synthesis methodology allows for the sequential application of thermal couplings, gradually transitioning from conventional designs to fully intensified systems such as DWCs. As illustrated in Fig. 8.7, these modifications include single and multiple thermal couplings, culminating in advanced configurations like DWC-R, which integrate rectification and depletion sections. These systems operate under optimized pressures to ensure energy efficiency (Van Duc Long & Lee, 2017, p. 1–210).

Water management is another critical aspect addressed in the intensified designs. The esterification reactions inherently produce water, reducing the need for additional freshwater streams. By integrating this water into the process, the designs minimize resource consumption and environmental impact, aligning with sustainable industrial practices (see Fig. 8.8) (Hallale, 2002).

These intensified configurations exemplify the principles of PI within a circular economy framework. By reducing energy consumption, optimizing resource use, and incorporating sustainable design principles, they provide innovative solutions to contemporary industrial challenges. The methodology demonstrates its effectiveness in generating advanced alternatives that combine economic, environmental, and operational benefits, ensuring their feasibility and appeal for industrial applications.

Process optimization

The development of sustainable chemical processes has significantly advanced over the past 5 decades, focusing on minimizing environmental impact, reducing energy consumption, lowering operational costs, and enhancing both safety and process operability. Achieving sustainable process designs presents notable challenges due to the complex interplay of nonlinearities, degrees of freedom, and large-scale multiobjective optimization problems inherent in chemical process models. For the purification of lactic acid via reactive distillation, this study targets optimizing process configurations to meet sustainability objectives. The selected metrics include economic performance, environmental impact, and safety. Economic performance is evaluated using the TAC, a metric that incorporates both capital and operating expenses, representing the financial viability of the process. Environmental impact is

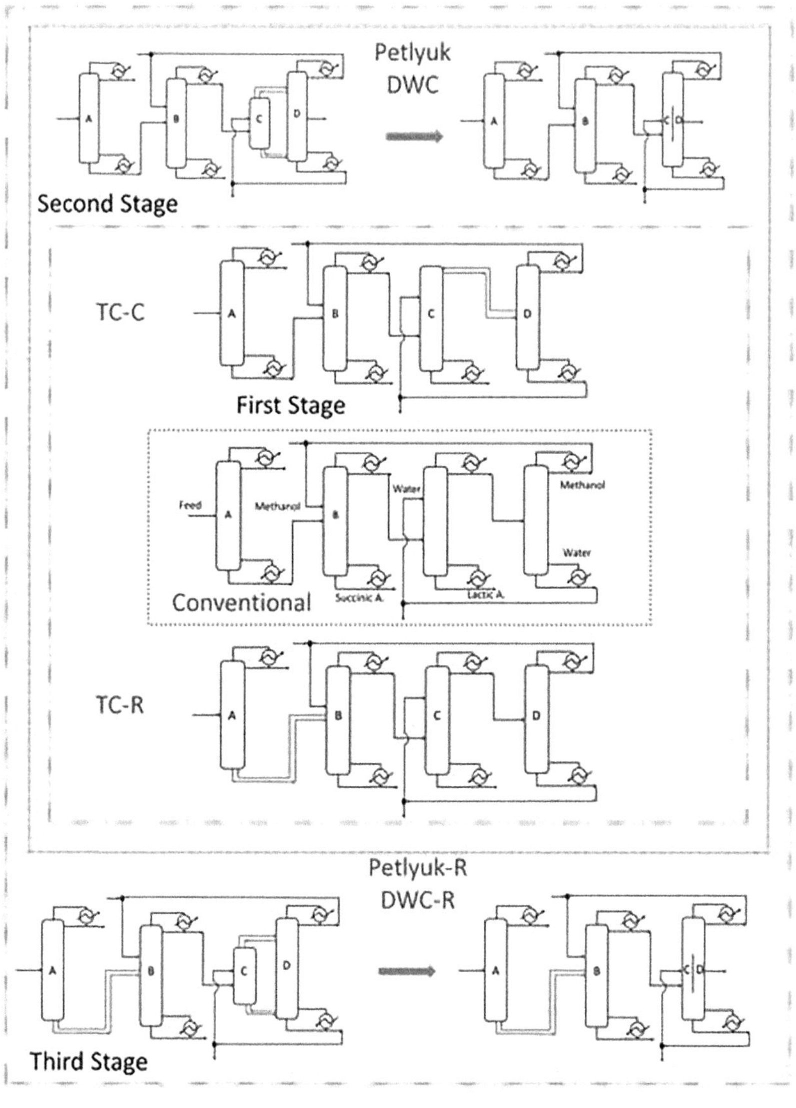

FIGURE 8.7 Intensification methodology applied to the conventional case.

quantified through the Eco-Indicator 99 (Eco-99), a life-cycle-based metric that measures the environmental burden, including resource depletion and ecosystem health. Safety is assessed using the individual risk (IR) index, which estimates the likelihood of injury or fatality for individuals exposed to hazards near the process. The objective function may be expressed as follow (Eq. 8.4):

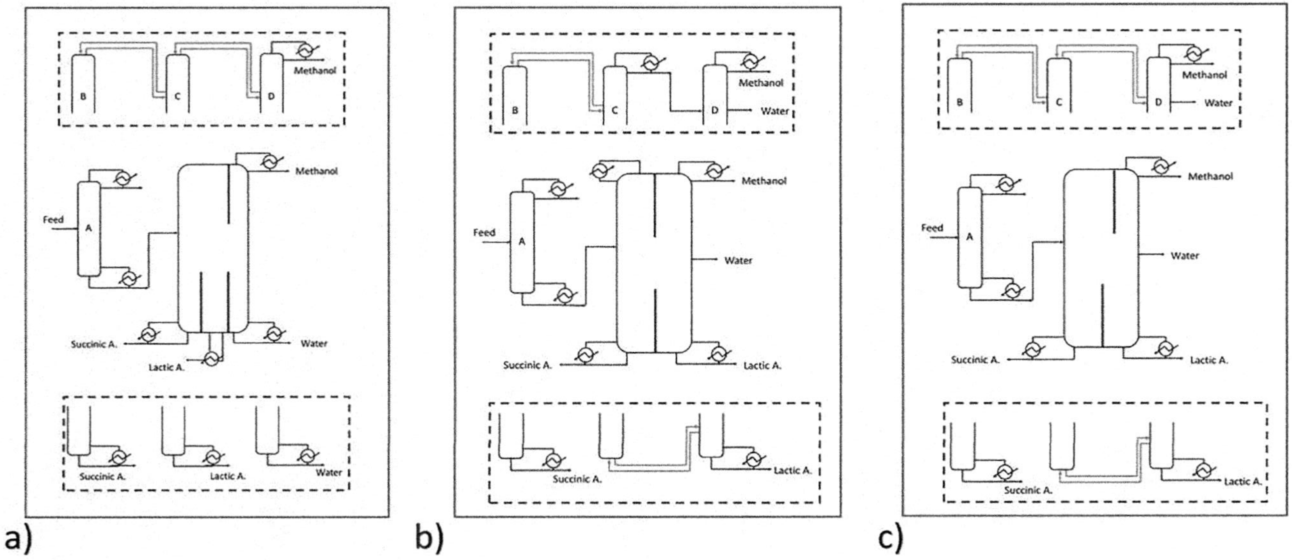

FIGURE 8.8 Generation of intensified alternatives with mass water integration DWC-A (a), DWC-B (b), and DWC-(c). All figures are presented in both expanded and DWC form.

$$\text{Min} \quad (\text{TAC}, \text{EI99}, \text{IR}) = f\left(N_t n, N_f n, R_r n, F_r n, F_l n, F_v n, D_{cn}, P_{cn}, FC_{cn}\right) \quad (8.4)$$

$$\text{Subject to} \quad \vec{x_m} > \vec{y_m} \quad (8.5)$$

y_m and x_m are the vectors of both obtained and required purities for the m_{th} components, respectively. Additionally, N_{tn} are total number of column stages, N_{fn} is the feed stage in column, R_{rn} is the reflux ratio, F_{rn} is the distillate/bottoms flux, F_{ln} is the interconnection liquid flow, F_{vn} is the interconnection vapor flow, D_{cn} is the column diameter, moreover for IR calculation many physicochemical properties are considered such as molecular weight, heat of combustion, LC_{50} and so on. The optimization process aims to simultaneously minimize these metrics, ensuring compliance with market standards by maintaining product purities above required thresholds. This multi-objective approach provides a comprehensive evaluation of the intensified configurations under consideration. The optimization methodology employs a hybrid metaheuristic method combining DETL enhancements. This hybrid approach, based on natural selection principles, improves computational efficiency and convergence by preventing repetitive exploration of previously visited solutions. The algorithm begins with an initial population of potential solutions, followed by iterative steps of mutation and crossover to generate diversity and refine solutions. Selection mechanisms prioritize the best-performing solutions while leveraging the tabu list to enhance solution space exploration. The iterative process continues over 900 generations, refining the results to identify optimal configurations. Dynamic data exchange between Aspen Plus and the optimization algorithm facilitates seamless evaluation of design variables. Input vectors are generated by the optimization algorithm and evaluated by Aspen Plus, which returns output vectors to guide subsequent iterations. Parameters such as population size, number of generations, tabu list proportion, crossover probability, and tabu radius were fine-tuned during preliminary calculations to ensure robust performance and convergence. This optimization framework provides a robust approach for identifying sustainable designs for lactic acid purification using reactive distillation. By minimizing economic, environmental, and safety metrics while ensuring operational controllability, it enables the development of efficient, cost-effective, and environmentally responsible chemical processes.

Results

This section presents the key findings from the optimization process for lactic acid purification using intensified separation schemes. The analysis focused on evaluating economic, environmental, and safety performance, complemented by an assessment of dynamic control properties. Purity constraints were established based on commercial requirements: 90% wt for lactic acid (aligned with the food and pharmaceutical industries), 99.9% wt for succinic acid, and 99.9% wt for methanol and water intended for recycling. These benchmarks reflect typical industrial specifications for these compounds (Rubio et al.,

2009). The optimization process involved 108,000 evaluations (120 individuals over 200 generations), yielding a clear identification of trends and optimal configurations. Three key performance indicators—TAC, eco-indicator 99 (EI99), and individual risk (IR)—were used to evaluate each design. Although all three indicators were optimized simultaneously, results are presented in two-dimensional graphs to enhance clarity (see Fig. 8.9).

The introduction of thermal couplings significantly improved economic and environmental performance, albeit with varying impacts on safety:

Stage 1: Initial Thermal Coupling

Thermal coupling replacing condensers reduced energy consumption, leading to TAC savings of up to 17%. In contrast, replacing reboilers generated only 11% savings and increased inherent safety risks slightly (approximately 1%) due to greater material exchange between columns. Energy reductions were more pronounced in configurations where the condenser was replaced, emphasizing the importance of targeting high-energy equipment for thermal coupling.

Stage 2: DWC Integration

While the DWC configuration eliminated the need for a reboiler, the resulting economic savings were marginal due to increased energy demands in other sections of the column. Environmental performance showed a slight decline (3% increase in EI99) compared to simpler thermal coupling designs. However, inherent safety improved, as fewer reboilers reduced material hold-ups, lowering accident probabilities by 10%.

Stage 3: DWC with Additional Thermal Coupling (DWC-R)

The DWC-R configuration eliminated reboilers in two columns but offered no significant economic or environmental advantages over the DWC. On the contrary, inherent safety decreased due to the increased energy intensity of the separation process (see parameters design in Tables 8.11–8.13).

The DWC-A alternative emerged as the most promising intensified design, demonstrating improvements in TAC (4% reduction) and inherent safety (13% reduction). This configuration allowed the integration of water produced during the esterification process into subsequent hydrolysis, eliminating the need for an external water feed. The annual water savings of approximately 2,143,819 kg translate to sufficient water for the basic consumption needs of 60 people annually, emphasizing its alignment with circular economy principles. Despite its advantages, the DWC-A configuration presented a penalty in EI99 due to higher energy requirements, a trade-off partially offset by capital cost reductions. The reuse of water and methanol streams further contributed to economic and environmental benefits, showcasing the potential of intensified designs to reduce resource consumption (see Table 8.14).

Dynamic performance varied significantly among configurations. Generally, alternatives with higher degrees of intensification displayed better controllability. Thermal couplings between the first two columns significantly enhanced dynamic stability. However, the integration of water streams tended

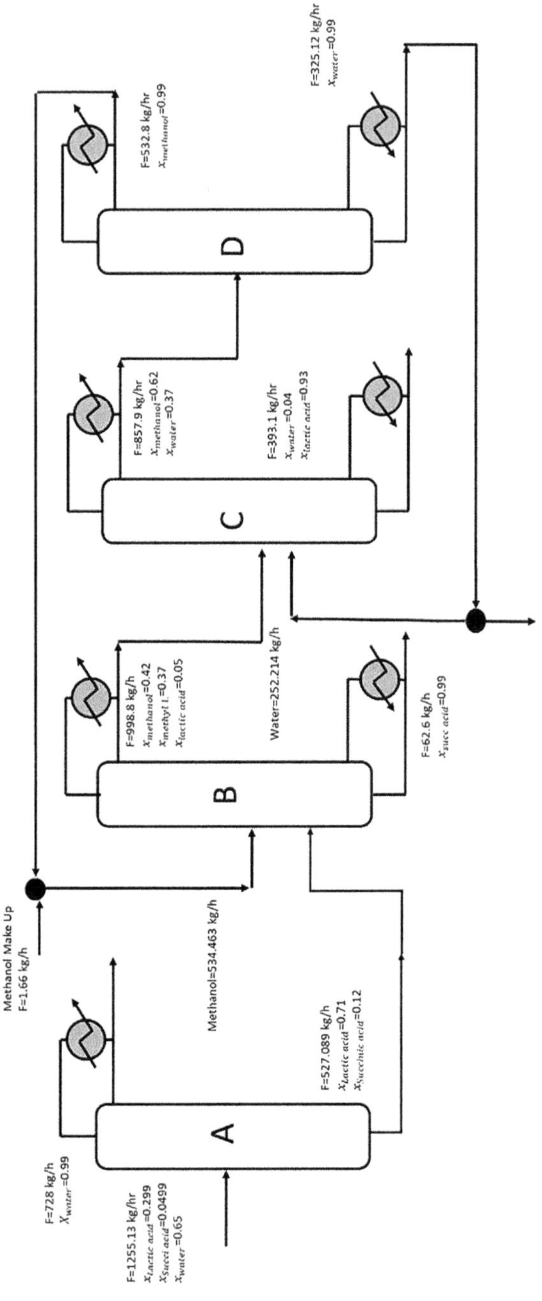

FIGURE 8.9 Mass balance for the conventional scheme, columns a–d.

TABLE 8.11 Design parameters of the conventional TC-C and TC-R schemes.

	Conventional				TC-C				TC-R			
Columns	C1	C2	C3	C4	C1	C2	C3	C4	C1	C2	C3	C4
Stages	43	8	60	32	42	8	59	31	46	8	60	31
Feed stage	23	4, 4	42, 42	14	24	5, 4	41, 46	17	15	4, 4	40, 34	23
Side stage								16	46			
Reflux ratio	0.1	0.05	0.12	1.82	0.04	0.026	–	1.77		0.29	0.05	0.9
Distillate flow (kmol/h)	40.4153	25.734	34.673	16.64	40.41	25.73		16.73	40.4	25.7	34.7	16.6
Bottoms flow (kmol/h	9.58	0.53	5.05	18.03	9.58	0.53	5.05	18.03	–	0.53	5.05	9.58
Interconnection flow (kmol/h)							5.899		14.5858			
Diameter (m)	1.03	1	0.86	1.02	1.03	1	0.86	1.02	1.03	1	0.86	1.02
Reactive stages		3–7	28–47			3–7	27–56			2–7	27–48	
Holdup (l)		29.14	97.062			28.9034124	91.6844112			30.5	96.6	
Pressure (atm)	0.1902	0.4599			0.211	0.37			0.12	0.37		
Reboiler duty (cal/s)	448,115				338,868				417,337			

Continued

TABLE 8.11 Design parameters of the conventional TC-C and TC-R schemes.—cont'd

	Conventional				TC-C				TC-R			
Succinic acid (% wt)		0.999				0.999				0.99		
Lactic acid (%wt)			0.94				0.914				0.95	
Methanol (%wt)				0.999				0.995				0.99
TAC ($/y)	1,274,559				1,062,454				1,134,718			
EI99 (eco-points/y)	405,025				302,771				376,834			
IR (probability/y)	0.000818084				0.000818023				0.000823006			

TABLE 8.12 Design parameters of the DWC and DWC-R schemes.

	DWC				DWC-R			
Columns	C1	C2	C3	C4	C1	C2	C3	C4
Stages	43	9	60	32	44	9	60	32
Feed stage	20	5, 7	28, 52	7, 25	15	4, 4	40, 34	23
Vapor interconnection flow (kg)			36.75		14.5858			
Input stage (vapor)			76		46			
Side stream (vapor)				25		5, 25		21
Side stream (liquid)				6		6		6
Side stream (water)				21		21		17
Reflux ratio	0.1	0.05		1.8518	0.1	0.29	0.05	0.9
Distillate flow (kmol/h)	40.4	25.7		16.65	40.4	25.7	34.7	16.6
Bottoms flow (kmol/h)	9.58	0.53	–	5.04	–	0.53	–	5.08
Diámetro (m)	1.03	1	0.86	1.02	1.03	1	0.86	1.02
Reactive stages		4—7	26–56			2–7	27–44	
Holdup (l)		42.37	90.07			30.5	96.6	
Pressure (atm)	0.194	0.569	1	1	0.12	0.37	1	1
Reboiler duty (cal/s)	349,927				338,292			

Continued

TABLE 8.12 Design parameters of the DWC and DWC-R schemes.—cont'd

	DWC				DWC-R			
Succinic acid (%wt)		0.999				0.99		
Lactic acid (%wt)			0.958				0.95	
Methanol (%wt)				0.999				0.99
TAC ($/y)	1,086,548				1,062,449			
EI99 (eco-points/y)	313,189				303,062			
IR (probability/y)	0.000734825				0.000736675			

TABLE 8.13 Design parameters of the thermally coupled sequence with water integration.

	TC-IM			
Columns	C1	C2	C3	C4
Stages	43	9	76	26
Feed stage	20	4, 4	–	7, 25
Vapor interconnection flow (kg)			36.75	
Input stage (vapor)			53	
Interconnection flow (vapor kmol/h)		28.53		25
Interconnection flow (liquid kmol/h)			2.8	6
Side stream (water)				21
Reflux ratio	0.1	–		2.095
Distillate flow (kmol/h)	40.41	–		16.682
Bottoms flow (kmol/h)	9.58	0.53	–	4.873
Diámetro (m)	1.03	1	0.86	1.02
Reactive stages		4–7	26–66	
Holdup (l)		42.37	90.07	
Pressure (atm)	0.19004	0.5331	1	1
Reboiler duty (cal/s)	387,370			
Succinic acid (%wt)		0.999		
Lactic acid (%wt)			0.958	
Methanol (%wt)				0.999
TAC ($/y)	1,111,585			
EI99 (eco-points/y)	365,591			
IR (probability/y)	0.000818369			

to penalize control properties, with the exception of the DWC-C scheme, which exhibited superior control due to its high level of energy integration. The DWC-A configuration, while optimal in terms of economic and environmental performance, faced challenges in control under closed-loop conditions. Conversely, simpler alternatives with fewer thermal couplings showed better controllability but were less economically and environmentally viable. The DWC-A design demonstrated the best balance of sustainability metrics, achieving significant savings in water consumption and operational costs while

TABLE 8.14 Design parameters of the DWC sequence with water integration.

	DWC-A				DWC-B				DWC-C			
Columns	C1	Wall A	Wall B	Main	C1	WALL A	wall B	Main	C1	Wall A	Wall B	Main
Stages	43	9	60	32	42	9	59	31	42	9	60	31
Feed stage	23	4, 4	1, 52	20	24	4, 4	1, 52	16, 18	15	4, 4	1, 52, 60	20,31
Side stage								15				14
Reflux ratio	0.085	–	–	1.88	0.0805	–	1.596	1.872		–	–	10.06
Distillate flow (kmol/h)	40.4106	–	–	16.66	40.4143	–		16.496	40.41	–	–	16.506
Bottoms flow (kmol/h	9.5893	0.531	6.09	2.985	9.587	0.5319	–	4.239	–	0.528	–	4.22
Interconnection flow (kmol/h)		28.637	2.899/ 59.654	40.008		28.535	2.89/ 23.61		14.5858	28.462	2.71/ 59.86/ 27.67	42.19/ 19.59
Diameter (m)	1.017	1.16	1.008	1.02	1.017	1.07	1.0183	1.03	1.017	1.1088	1.014	1.08
Reactive stages		3–7	28–47			3–7	27–47			2–7	27–48	
Holdup (l)		29.14	97.062			28.903	91.684			30.5	96.6	
Pressure (atm)	0.189	0.4599	1	1	0.181	0.37	1	1	0.18	0.37	1	1

Reboiler duty (cal/s)	391,938				536,473				1,447,227			
Succinic acid (%wt)		0.999				0.999				0.99		
Lactic acid (% wt)			0.939					0.914				0.95
Methanol (%wt)				0.999				0.995				0.99
TAC ($/y)	1,044,799				1,498,392				2,871,129			
EI99 (eco-points/y)	356,249				500,695				963,497			
IR (probability/y)	0.000652447				0.000652398				0.000652398			

improving safety. However, its complex topology posed control challenges, highlighting a trade-off between intensification and operability. This underscores the need for a holistic approach to process design, considering not only sustainability indicators but also operational feasibility (González-Navarrete et al., 2022).

The analysis of various lactic acid separation and purification schemes revealed critical insights into the performance of intensified configurations. Replacing reboilers with thermal couplings showed minimal reductions in environmental and economic impacts and slightly increased the inherent risk of the schemes. Conversely, replacing condensers with thermal couplings consistently improved performance indicators, highlighting their greater effectiveness in process optimization. However, schemes that substituted reboilers generally exhibited a compromise in performance metrics.

Among all the evaluated designs, the DWC-A scheme emerged as the most promising, achieving approximately 4% savings in total annual cost (TAC) and a 12% reduction in individual risk (IR). These improvements underscore the economic and safety advantages of this configuration. A noteworthy trend was observed in equipment sizing, where larger equipment, while increasing capital costs and environmental impact due to steel usage, also led to higher IR values because of the greater volume of material inside the columns. Comparing alternatives with and without mass integration, those incorporating mass integration demonstrated better economic and environmental performance. The DWC-A scheme stood out in this regard, offering the lowest impact among the alternatives. Nonetheless, a preliminary controllability assessment indicated that simpler topological configurations did not guarantee superior control properties. Mass-integrated schemes generally provided enhanced controllability but at the expense of higher costs and environmental impact. These findings highlight the importance of balancing economic, environmental, and operational criteria in the design of intensified separation processes.

Conclusions

Chapter 8 provides a comprehensive exploration of the design, synthesis, and optimization of intensified separation processes, highlighting their potential to address some of the most pressing challenges in chemical process engineering. The chapter focused on two major case studies: the purification of biobutanol and lactic acid, showcasing the transformative role of PI strategies and DWC.

For the biobutanol case study, hybrid configurations combining liquid-liquid extraction with intensified DWC designs demonstrated remarkable improvements in economic and environmental metrics. The substitution of traditional column setups with intensified schemes achieved reductions in TAC by up to 22% and Eco-Indicator 99 (EI99) by 18%, while also providing savings in water consumption. These findings underscore the significant energy efficiency and resource conservation benefits of integrating PI strategies into bioproduct separations.

In the case of lactic acid purification, the chapter examined the use of reactive distillation and thermally coupled configurations to overcome the challenges posed by azeotropic behavior and dilute feed compositions. Among the configurations analyzed, the DWC-A scheme emerged as the most promising, achieving substantial improvements in TAC and inherent safety (IR). Additionally, the DWC designs allowed for the integration and reuse of recovered streams, such as water and methanol, further enhancing the sustainability of the process and aligning with circular economy principles.

A notable conclusion from this chapter is that while intensified configurations such as DWC and RD offer substantial benefits in terms of energy efficiency, cost reduction, and environmental impact, these advantages are often accompanied by trade-offs in controllability and operational complexity. The inherent safety of these designs also requires careful consideration, as the interaction of thermal couplings and internal flows can influence risk profiles. From a broader perspective, the findings reinforce the critical role of systematic design methodologies in identifying and optimizing PI configurations. The use of multiobjective optimization frameworks enabled the simultaneous evaluation of economic, environmental, and safety metrics, providing a balanced approach to sustainable process development.

Thus, Chapter 8 highlights the potential of PI technologies, such as DWC and RD, to revolutionize chemical separations by delivering sustainable, cost-effective, and energy-efficient solutions. The insights and methodologies presented in this chapter provide a valuable roadmap for the industrial implementation of intensified separation processes, addressing the growing demand for greener and more efficient chemical processes in a competitive global market.

References

Abdel-Rahman, M. A., Tashiro, Y., & Sonomoto, K. (2013). Recent advances in lactic acid production by microbial fermentation processes. *Biotechnology Advances, 31*(6), 877–902. https://doi.org/10.1016/j.biotechadv.2013.04.002

Amiri, H., & Karimi, K. (2015). Improvement of acetone, butanol, and ethanol production from woody biomass using organosolv pretreatment. *Bioprocess and Biosystems Engineering, 38* (10), 1959–1972. https://doi.org/10.1007/s00449-015-1437-0

Castillo-Landero, A., Jiménez-Gutiérrez, A., & Gani, R. (2018). Intensification methodology to minimize the number of pieces of equipment and its application to a process to produce dioxolane products. *Industrial and Engineering Chemistry Research, 57*(30), 9810–9820. https://doi.org/10.1021/acs.iecr.7b05229

Cavani, F., Albonetti, S., Basile, F., & Gandini, A. (2016). *Chemicals and fuels from bio-based building blocks*. Wiley-VCH.

Contreras-Vargas, C. A., Gómez-Castro, F. I., Sánchez-Ramírez, E., Segovia-Hernández, J. G., Morales-Rodríguez, R., & Gamiño-Arroyo, Z. (2019). Alternatives for the purification of the blend butanol/ethanol from an acetone/butanol/ethanol fermentation effluent. *Chemical Engineering and Technology, 42*(5), 1088–1100. https://doi.org/10.1002/ceat.201800641

Corma Canos, A., Iborra, S., & Velty, A. (2007). Chemical routes for the transformation of biomass into chemicals. *Chemical Reviews, 107*(6), 2411–2502. https://doi.org/10.1021/cr050989d

Crowl, D. A., & Louvar, J. F. (2001). *Chemical process safety: Fundamentals with applications*. Pearson Education.

Dejanović, I., Matijašević, L., Halvorsen, I. J., Skogestad, S., Jansen, H., Kaibel, B., & Olujić, Ž. (2011). Designing four-product dividing wall columns for separation of a multicomponent aromatics mixture. *Chemical Engineering Research and Design, 89*(8), 1155–1167. https://doi.org/10.1016/j.cherd.2011.01.016

Dejanović, I., Matijašević, L., & Olujić, Ž. (2010). Dividing wall column—A breakthrough towards sustainable distilling. *Chemical Engineering and Processing: Process Intensification, 49*(6), 559–580. https://doi.org/10.1016/j.cep.2010.04.001

Delgado-Delgado, R., Hernández, S., Barroso-Muñoz, F. O., Segovia-Hernández, J. G., & Castro-Montoya, A. J. (2012). From simulation studies to experimental tests in a reactive dividing wall distillation column. *Chemical Engineering Research and Design, 90*(7), 855–862. https://doi.org/10.1016/j.cherd.2011.10.019

Errico, M., Rong, B., Tola, G., & Turunen, I. (2009). A method for systematic synthesis of multicomponent distillation systems with less than N-1 columns. *Chemical Engineering and Processing: Process Intensification, 48*, 907–920. https://doi.org/10.1016/j.cep.2008.12.005

Errico, M., Sanchez-Ramirez, E., Quiroz-Ramìrez, J. J., Rong, B. G., & Segovia-Hernandez, J. G. (2017). Multiobjective optimal acetone-butanol-ethanol separation systems using liquid-liquid extraction-assisted divided wall columns. *Industrial and Engineering Chemistry Research, 56*(40), 11575–11583. https://doi.org/10.1021/acs.iecr.7b03078

Errico, M., Sanchez-Ramirez, E., Quiroz-Ramìrez, J. J., Segovia-Hernandez, J. G., & Rong, B. G. (2016). Synthesis and design of new hybrid configurations for biobutanol purification. *Computers and Chemical Engineering, 84*, 482–492. https://doi.org/10.1016/j.compchemeng.2015.10.009

Faraj, K., Khaled, M., Faraj, J., Hachem, F., & Castelain, C. (2021). A review on phase change materials for thermal energy storage in buildings: Heating and hybrid applications. *Journal of Energy Storage, 33*, Article 101913. https://doi.org/10.1016/j.est.2020.101913

Górak, A., & Schoenmakers, H. (2014). *Distillation: Operation and applications*. Elsevier Inc.. https://doi.org/10.1016/C2010-0-67739-X

Gadalla, M. A., Olujic, Z., Jansens, P. J., Jobson, M., & Smith, R. (2005). Reducing CO_2 emissions and energy consumption of heat-integrated distillation systems. *Environmental Science and Technology, 39*(17), 6860–6870. https://doi.org/10.1021/es049795q

González-Navarrete, C., Sánchez-Ramírez, E., Ramírez-Márquez, C., Hernández, S., Cossío-Vargas, E., & Segovia-Hernández, J. G. (2022). Innovative reactive distillation process for the sustainable purification of lactic acid. *Industrial and Engineering Chemistry Research, 61*(1), 621–637. https://doi.org/10.1021/acs.iecr.1c04050

Hallale, N. (2002). A new graphical targeting method for water minimisation. *Advances in Environmental Research, 6*(3), 377–390. https://doi.org/10.1016/s1093-0191(01)00116-2

Halvorsen, I. J., & Skogestad, S. (2003). Minimum energy consumption in multicomponent distillation. 3. More than three products and generalized Petlyuk arrangements. *Industrial and Engineering Chemistry Research, 42*(3), 616–629. https://doi.org/10.1021/ie0108651

Harvianto, G. R., Kim, K. H., Kang, K. J., & Lee, M. (2019). Optimal operation of a dividing wall column using an enhanced active vapor distributor. *Chemical Engineering Research and Design, 144*, 512–519. https://doi.org/10.1016/j.cherd.2019.02.038

Hernández, S. (2008). Analysis of energy-efficient complex distillation options to purify bioethanol. *Chemical Engineering and Technology, 31*(4), 597–603. https://doi.org/10.1002/ceat.200700467

Hernández, S., & Jiménez, A. (1999). Design of energy-efficient Petlyuk systems. *Computers and Chemical Engineering, 23*(8), 1005–1010. https://doi.org/10.1016/s0098-1354(99)00257-4

Ignat, R. M., & Kiss, A. A. (2013). Optimal design, dynamics and control of a reactive DWC for biodiesel production. *Chemical Engineering Research and Design, 91*(9), 1760–1767. https://doi.org/10.1016/j.cherd.2013.02.009

Jafari, Y., Amiri, H., & Karimi, K. (2016). Acetone pretreatment for improvement of acetone, butanol, and ethanol production from sweet sorghum bagasse. *Applied Energy, 168*, 216–225. https://doi.org/10.1016/j.apenergy.2016.01.090

Kaibel, G. (1987). Distillation columns with vertical partitions. *Chemical Engineering and Technology, 10*(1), 92–98. https://doi.org/10.1002/ceat.270100112

Karimi, K., Tabatabaei, M., Horváth, I. S., & Kumar, R. (2015). Recent trends in acetone, butanol, and ethanol (ABE) production. *Biofuel Research Journal, 2*(4), 301–308. https://doi.org/10.18331/BRJ2015.2.4.4

Kim, S. Y., Kim, D. M., & Lee, B. (2017). Process simulation for the recovery of lactic acid using thermally coupled distillation columns to mitigate the remixing effect. *Journal of Chemical Engineering, 34*(5), 1310–1318. https://doi.org/10.1007/s11814-017-0009-1

Kiss, A. A. (2013). *Advanced distillation technologies: Design, control and applications.* John Wiley and Sons. https://doi.org/10.1002/9781118543702

Komesu, A., Martins, P. F., Lunelli, B. H., Rocha, J. O., Maciel Filho, R., & Wolf Maciel, M. R. (2015). The effect of evaporator temperature on lactic acid purity and recovery by short path evaporation. *Separation Science and Technology, 50*(10), 1548–1553. https://doi.org/10.1080/01496395.2014.975363

Lestak, F., & Collins, C. (1997). Distillation for the 21st century. *Chemical Engineering, 104*(7), 72–76.

Lukač, G., Halvorsen, I. J., Olujić, Ž., & Dejanović, I. (2019). On controllability of a fully thermally coupled four-product dividing wall column. *Chemical Engineering Research and Design, 147*, 367–377. https://doi.org/10.1016/j.cherd.2019.04.041

Preißinger, U., Ränger, L.-M., & Grützner, T. (2019). Design considerations of a simplified multiple dividing wall column pilot plant. *ChemEngineering, 3*(2), 34. https://doi.org/10.3390/chemengineering3020034

Ramapriya, G. M., Tawarmalani, M., & Agrawal, R. (2016). Thermal coupling links to liquid-only transfer streams: An enumeration method for new FTC dividing wall columns. *AIChE Journal, 62*(4), 1200–1211. https://doi.org/10.1002/aic.15053

Rangaiah, G., & Pandu. (2010). *Stochastic global optimization : Techniques and applications in chemical engineering*. World Scientific.

Rong, B.-G. (2011). Synthesis of dividing-wall columns (DWC) for multicomponent distillations—A systematic approach. *Chemical Engineering Research and Design, 89*(8), 1281–1294. https://doi.org/10.1016/j.cherd.2011.03.014

Rong, B.-G., Kraslawski, A., & Turunen, I. (2004). Synthesis and optimal design of thermodynamically equivalent thermally coupled distillation systems. *Industrial and Engineering Chemistry Research, 43*(18), 5904–5915. https://doi.org/10.1021/ie0341193

Rubio, O. A. P., Jørgensen, S. B., & Jonsson, G. E. (2009). Lactic acid recovery in electro-enhanced dialysis: Modelling and validation. *Computer Aided Chemical Engineering, 26*, 773–778.

Sánchez-Ramírez, E., Quiroz-Ramírez, J. J., Segovia-Hernández, J. G., Hernández, S., & Bonilla-Petriciolet, A. (2015). Process alternatives for biobutanol purification: design and optimization. *Industrial & Engineering Chemistry Research, 54*(1), 351–358.

Schultz, M. A., Stewart, D. G., Harris, J. M., Rosenblum, S. P., Shakur, M. S., & Brien, D. (2002). *Reduce costs with dividing wall columns.*

Segovia-Hernández, J. G., Hernández, S., Sánchez-Ramírez, E., & Mendoza-Pedroza, J. (2021). A short review of dividing wall distillation column as an application of process intensification: Geographical development and the pioneering contribution of Prof. Arturo Jimenez in Latin America. *Chemical Engineering and Processing - Process Intensification, 160*, Article 108275. https://doi.org/10.1016/j.cep.2020.108275

Soraya Rawlings, E., Chen, Q., Grossmann, I. E., & Caballero, J. A. (2019). Kaibel column: Modeling, optimization, and conceptual design of multi-product dividing wall columns. *Computers and Chemical Engineering, 125*, 31–39. https://doi.org/10.1016/j.compchemeng.2019.03.006

Sotudeh, N., & Shahraki, B. H. (2007). A method for the design of divided wall columns. *Chemical Engineering and Technology, 30*(9), 1284–1291. https://doi.org/10.1002/ceat.200700175

Van Duc Long, N., & Lee, M. (2017). *Advances in distillation retrofit advances in distillation Retrofit.* Jordan: Jordan Springer Singapore. https://doi.org/10.1007/978-981-10-5901-8

Wolff, E. A., & Skogestad, S. (1995). Operation of integrated three-product (Petlyuk) distillation columns. *Industrial and Engineering Chemistry Research, 34*(6), 2094–2103. https://doi.org/10.1021/ie00045a018

Yildirim, O., Kiss, A. A., & Kenig, E. Y. (2011). Dividing wall columns in chemical process industry: A review on current activities. *Separation and Purification Technology, 80*(3), 403–417. https://doi.org/10.1016/j.seppur.2011.05.009

Chapter 9

Innovations, AI applications and other environmentally sustainable applications

Introduction

The digital transformation and automation of chemical processes through the application of neural networks signify a paradigm shift toward achieving sustainability, aligning with the United Nations' 2030 Agenda for Sustainable Development. By integrating these advanced computational tools, industries can enhance resource utilization, minimize waste generation, and mitigate emissions, all while advancing process optimization and enabling data-driven decision-making. These technological advancements are reshaping the chemical industry, particularly by enabling the processing and analysis of vast real-time datasets. This capability enhances operational efficiency by optimizing the use of raw materials and energy, thereby reducing waste and emissions (Biermann et al., 2017). This aligns directly with Sustainable Development Goal (SDG) 12, which emphasizes responsible consumption and production.

Moreover, the integration of neural networks into process optimization frameworks significantly improves product quality, ensuring consistency and safety in chemical production. By enhancing operational efficiency and technological innovation, these advancements directly contribute to SDG 9 (Industry, Innovation, and Infrastructure). A critical facet of this transformation is enhanced safety within the chemical sector. Neural networks' predictive and proactive capabilities enable early detection of potential issues, significantly reducing risks associated with accidents, spills, and hazardous workplace conditions. This directly supports SDG 8 (Decent Work and Economic Growth) by fostering safer, more sustainable work environments.

Furthermore, the ability of neural networks to analyze and adapt to real-time operational data reduces costs and carbon footprints, supporting SDG 13 (Climate Action) by lowering greenhouse gas emissions and conserving resources (Lu et al., 2015). In this chapter, we will explore how digitization

Thermally Coupled Distillation Columns. https://doi.org/10.1016/B978-0-443-33114-5.00010-7

and automation, driven by neural networks, are transforming the chemical industry by addressing economic, environmental, and social challenges in alignment with the 2030 Agenda.

Neural networks in biofuel production

The adoption of digital technologies, particularly neural networks, is especially relevant in the production of biofuels, a cornerstone of sustainable energy systems. Biofuels, derived from renewable biomass and agricultural products, represent a critical pathway for reducing greenhouse gas emissions and achieving energy security. Neural networks facilitate the optimization of the intricate processes involved in biofuel production, including feedstock conversion, energy use, and quality control. By doing so, these technologies advance SDG 7 (Affordable and Clean Energy) and SDG 15 (Life on Land) through sustainable land use practices for biomass production (Ogunmakinde et al., 2022).

Biofuel production processes often involve complex enzymatic and microbial reactions, requiring precise real-time control to ensure stability, prevent contamination, and maximize efficiency. Neural networks enable predictive maintenance and dynamic process adjustments, ensuring robust operational performance. The ability of these technologies to address both technical and sustainability challenges underscores their role in advancing global energy transitions. For example, biofuels such as ethanol and butanol, derived from biomass fermentation, are characterized by low emissions and compatibility with existing energy systems (Ribeiro et al., 2007).

Intensified processes in biofuel purification

The integration of intensified processes, such as dividing wall columns (DWC), represents a significant advancement in biofuel purification. DWCs enable the separation of complex mixtures in a single column, reducing energy and capital costs compared to conventional distillation methods (Gómez-Castro et al., 2008). This aligns with the principles of process intensification, which emphasize reducing equipment size, energy consumption, and environmental impacts. Dynamic properties, such as controllability, are critical to the design of DWCs. While their complexity poses challenges, innovative control strategies have demonstrated their potential to achieve substantial cost and energy savings (Rewagad & Kiss, 2012).

DWCs offer a transformative solution for separating bioproducts such as vegetable oils and organic materials for biofuel production. This technology not only supports the transition to renewable energy but also aligns with the UN's 2030 Agenda by addressing global challenges such as resource scarcity, energy security, and environmental protection. The exploration of these technologies in this chapter highlights their potential to contribute to a more sustainable and equitable future.

Industry 4.0 and machine learning in chemical processes

The convergence of Industry 4.0 technologies and machine learning algorithms plays a pivotal role in advancing the chemical industry toward sustainability. Industry 4.0, characterized by the integration of automation, digitization, and data-centric operations, enables optimized resource allocation, reduced environmental impacts, and enhanced operational efficiency. Machine learning, as a subset of artificial intelligence (AI), enhances this potential through predictive analytics, pattern recognition, and adaptive optimization, addressing challenges across diverse sectors including energy and agriculture.

In chemical engineering, the application of AI algorithms is rapidly advancing, driven by increased computational capabilities and data availability. Among AI techniques, artificial neural networks (ANNs) stand out for their ability to model highly nonlinear systems. For instance, ANNs have been utilized in conjunction with model predictive control to enhance distillation column performance (Shin et al., 2020). While these applications have demonstrated feasibility, challenges remain in fine-tuning ANN architectures for optimal performance. This chapter outlines a systematic methodology for designing ANN architectures tailored to complex chemical processes, exemplified by biofuel production systems.

Challenges and opportunities

Despite their potential, the application of neural networks in chemical engineering has predominantly relied on black-box computational packages. While these tools offer convenience, they often limit transparency and customization, impacting performance. This work emphasizes the use of open-source platforms to enable direct interaction with neural network architectures, enhancing their design and applicability.

The innovation of digitizing and automating biofuel purification processes through neural networks represents a significant contribution to the fields of sustainable energy and chemical engineering. Neural networks enable dynamic, real-time adaptability to changes in feedstock composition, temperature, pressure, and other operational variables. This capability ensures efficient biofuel purification while meeting stringent quality and sustainability standards. By addressing these challenges, this chapter highlights the transformative potential of neural networks in fostering sustainable development in the chemical industry.

Methodology to determine neural network architectures

This chapter presents a robust methodology to forecast the dynamics of complex distillation columns using ANNs. By leveraging statistical principles, we optimize the ANN architecture to accurately model the dynamic behavior of such systems. The methodology incorporates key aspects such as activation functions, training algorithms, the number of hidden layers, and neurons per layer. Each step is described in detail to ensure reproducibility and facilitate its application to similar systems.

Step 1: Data collection

To demonstrate the feasibility and efficacy of the proposed methodology, Aspen Dynamics was utilized to model the intricate dynamics of a divided-wall distillation column (DWC). The DWC configuration, as detailed in the seminal work of Errico et al. (2017), was employed to separate a ternary mixture of acetone, butanol, and ethanol (ABE). This system was chosen due to its complexity and relevance to industrial applications.

The feed stream to the DWC consisted of acetone, butanol, ethanol, and water in weight fractions of 0.3018, 0.1695, 0.0073, and 0.5214, respectively. To achieve water removal, n-hexyl acetate was employed as the solvent in a liquid-liquid extraction process, followed by water recovery using conventional distillation. Critical operational variables—reflux ratio, reboiler duty, and side flowrate—were systematically manipulated to study their impact on the separation dynamics.

Dynamic simulations were performed in an open-loop configuration, introducing perturbations to the variables based on established guidelines (Luyben, 1992). Perturbations ranged from $\pm 5\%$ and were applied continuously to simulate realistic process variations. Data acquisition was conducted at 5-hour intervals over a 100-hour timeline, capturing the dynamic responses of the ABE components. This comprehensive dataset forms the foundation for ANN training and validation.

Remark: To ensure the ANN adequately estimates parameters, the concept of "persistent excitation" (Narendra & Annaswamy, 2007, p. 127–160) was incorporated. This approach ensures that network weights are consistently stimulated, enabling robust parameter estimation in both mechanistic and black-box models (Nar & Sastry, 2019). Data was collected in a closed-loop configuration using PI controllers, with simultaneous tuning of three set-points to generate time-varying operational datasets.

Step 2: Data visualization

Prior to applying machine learning tools, thorough data visualization is critical to uncover potential correlations and inconsistencies. Correlation matrices and scatter plots are particularly effective in identifying relationships between features and target variables.

Using Python's Seaborn library, the data generated in Aspen Dynamics was visualized. For instance, Fig. 9.1a highlights that the reflux ratio and reboiler duty significantly influence the butanol composition profile, while the side flowrate introduces a more complex dynamic. Fig. 9.1b reveals that butanol and acetone exhibit similar dynamic behaviors with respect to these variables, whereas ethanol demonstrates a distinct correlation with the side stream flowrate. These insights guide the selection of features for ANN training.

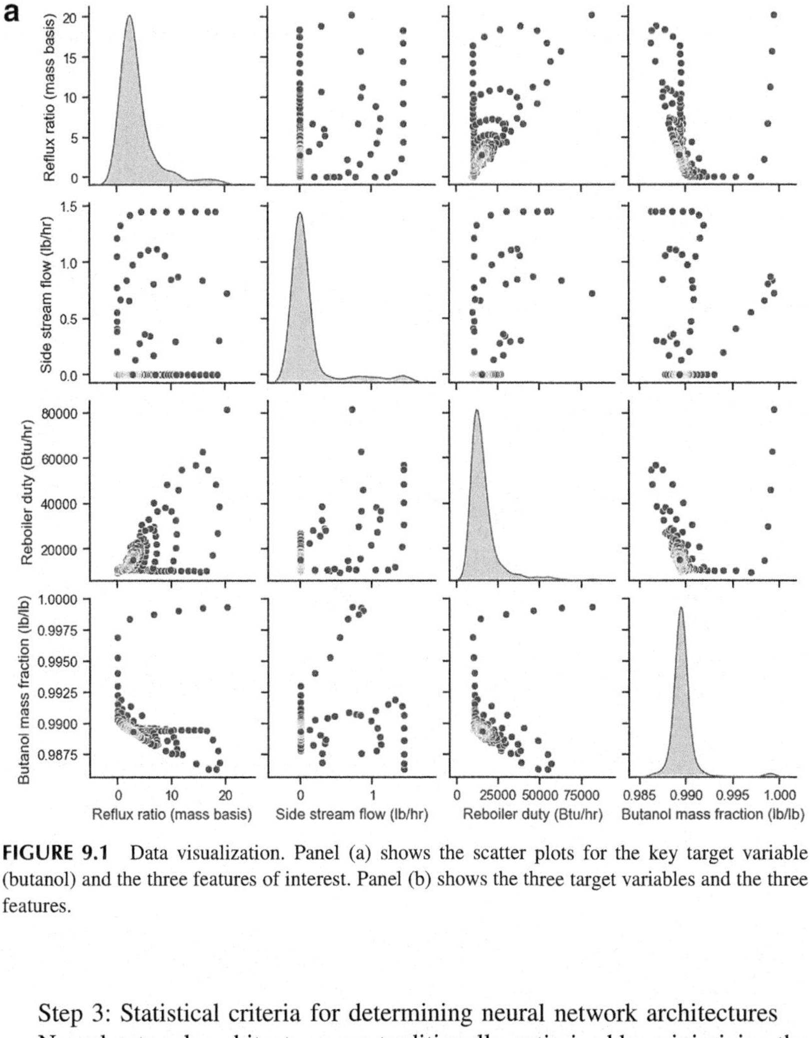

FIGURE 9.1 Data visualization. Panel (a) shows the scatter plots for the key target variable (butanol) and the three features of interest. Panel (b) shows the three target variables and the three features.

Step 3: Statistical criteria for determining neural network architectures

Neural network architectures are traditionally optimized by minimizing the mean squared error (MSE) (Biyanto et al., 2010; Bulsari & Palosaafi, 1993; Shin et al., 2020). While effective, focusing solely on MSE may lead to overfitting due to increased model complexity. To address this, we employ Akaike's Information Criterion (AIC) as a model selection tool. AIC balances model goodness-of-fit with complexity, providing a robust framework for selecting the optimal ANN architecture (Cavanaugh & Neath, 2019).

The AIC is calculated as follows:

$$\mathrm{AIC} = \mathrm{N}\log(\mathrm{MSE}) + (2\mathrm{MN}/(\mathrm{N}-\mathrm{M}-1))$$

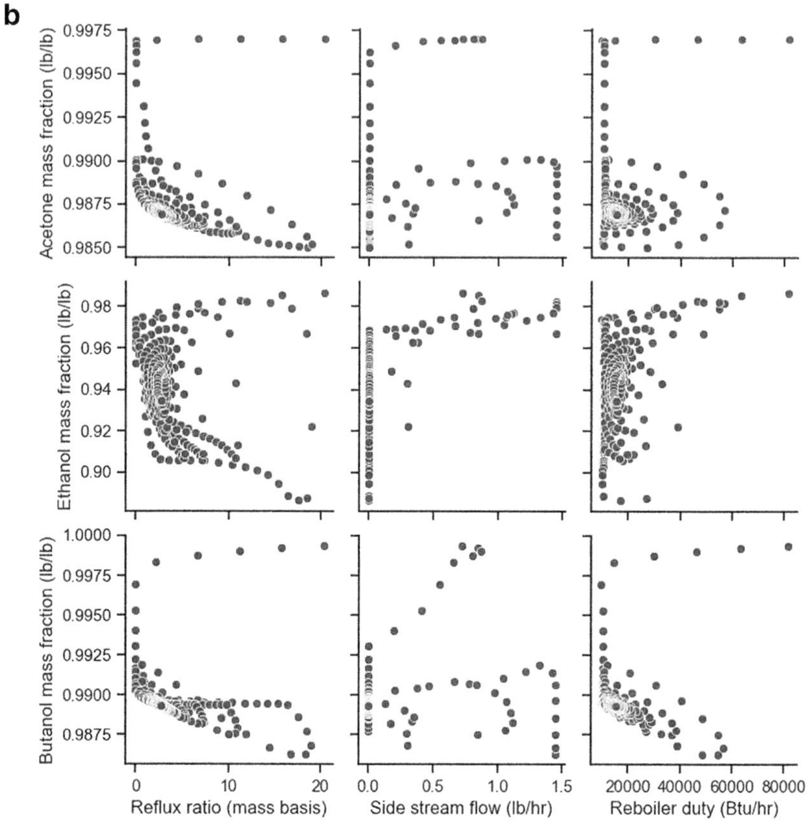

FIGURE 9.1 cont'd

where M represents the number of unknown parameters (e.g., ANN weights), N is the number of data points, and the mean square error is the MSE. Lower AIC values indicate better model performance.

Step 4: Data normalization

Data normalization is a critical preprocessing step for ANN training. All feature values were scaled to the [0,1] interval, ensuring consistent input magnitudes and improved training stability.

Step 5: Model training

The normalized dataset was split into training (76%) and testing (24%) subsets. A single hidden-layer ANN was trained to predict the dynamic behavior of the DWC. Fig. 9.2 illustrates the reduction in MSE as the number of training epochs increased, converging to a minimum value. This process was used to determine the optimal number of epochs for the final model.

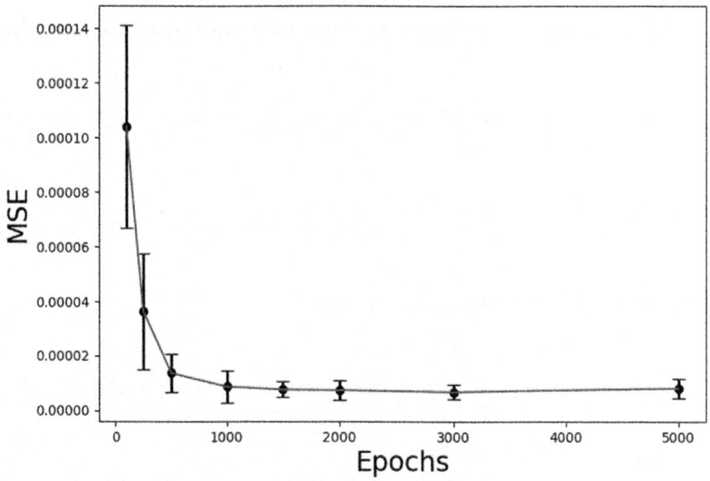

FIGURE 9.2 Training process.

Step 6: Exploring neural network parameters and architectures

Using the AIC framework, we systematically evaluated various ANN configurations to identify a minimalistic yet effective architecture. Parameters explored included:

Number of hidden layers and neurons per layer: A single hidden layer was found sufficient for this case study.

Activation functions: Sigmoid, ReLU, and tanh functions were evaluated for their impact on model performance.

Training algorithms: Gradient descent, Adam, and RMSprop optimizers were compared.

Fig. 9.3 presents the selected architecture, demonstrating its predictive accuracy and computational efficiency. By optimizing these parameters, the ANN effectively modeled the DWC dynamics, providing a valuable tool for process control and optimization.

The methodology outlined in this chapter provides a systematic approach to designing ANN architectures for complex process modeling. By combining statistical criteria, rigorous data preprocessing, and targeted parameter exploration, the proposed framework ensures robust and interpretable results. This approach is readily adaptable to other chemical processes, offering a scalable solution for industrial applications.

Case study in biobutanol purification

The production of biobutanol, an important biofuel, can be achieved through the ABE fermentation process, utilizing renewable biomass sources such as

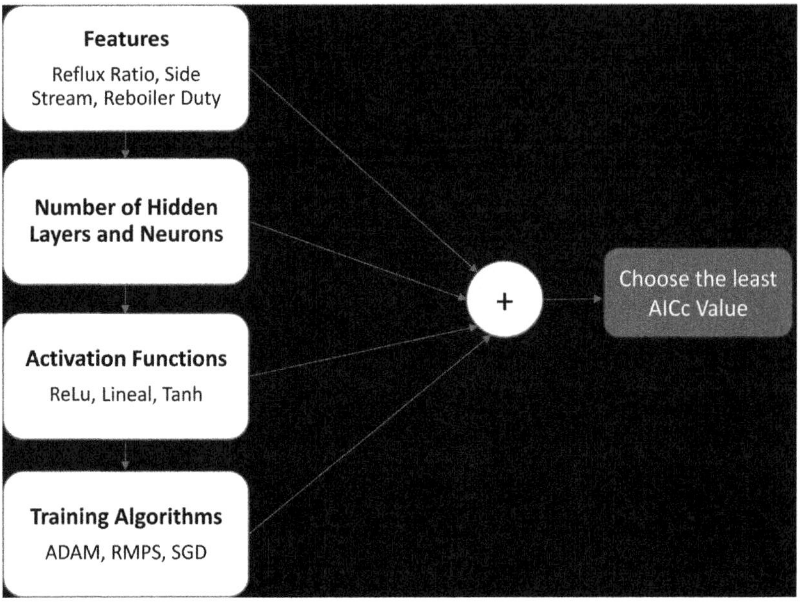

FIGURE 9.3 Methodology for determining a neural network architecture.

lignocellulose. This fermentation results in a complex mixture of acetone, ethanol, and butanol, which presents significant separation challenges due to the overlapping boiling points and the energy-intensive nature of traditional separation techniques. Biobutanol, when used as a fuel in blends, offers a clear advantage over fossil-derived butanol, as it is produced from renewable sources, thus emphasizing its environmental benefits despite sharing the same chemical properties with fossil-derived counterparts. To overcome the separation difficulties inherent in biobutanol production, Errico et al. (2017) proposed intensified separation alternatives that combine liquid-liquid extraction columns with conventional distillation columns, creating a more efficient system.

In the conventional approach, the separation of acetone, ethanol, and butanol occurs in multiple stages. Initially, acetone is separated as the distillate in the first distillation column, leaving a mixture of ethanol and butanol in the bottoms product. This mixture is then fed into a second column, where ethanol is distilled out, and butanol remains in the bottoms product. However, to improve energy efficiency, DWC offer a promising alternative. In this intensified configuration, the DWC efficiently separates the ternary ABE mixture into acetone, ethanol, and butanol by adjusting key operational parameters such as reflux ratio, reboiler duty, and lateral outflow. The DWC provides acetone as the top product, ethanol through the side outlet, and butanol as the bottom product, all with reduced energy consumption compared to conventional methods.

Simulations of these separation processes were carried out using Aspen Dynamics, with the NRTL-HOC thermodynamic model employed to describe phase equilibrium. The simulation's input data for the liquid-liquid extraction column comprised weight fractions of acetone (0.3018), butanol (0.1695), ethanol (0.0073), and water (0.5214). N-hexyl acetate was used as the extraction solvent to selectively separate water, which was later recovered using conventional distillation. The data from these simulations were utilized to train ANNs designed to model the process and optimize the separation parameters.

Neural network configuration and optimization

The ANN approach aimed at minimizing computational complexity while ensuring high predictive accuracy. A variety of network architectures and optimization algorithms, including ADAM, RMSPROP, and Stochastic Gradient Descent (SGD), were explored. The performance of each configuration was assessed using the AIC to determine the optimal number of neurons. For the ADAM algorithm, the Tanh activation function required 10 neurons, the Linear function needed five neurons, and the ReLU function was best with 15 neurons. For RMSPROP, Tanh performed well with 15 neurons, whereas Linear and ReLU functions performed optimally with five neurons each. Similarly, SGD showed optimal performance with five neurons for all activation functions.

A comparative analysis of the three activation functions—Linear, ReLU, and Tanh—revealed that configurations with more than five hidden neurons did not lead to substantial reductions in MSE. Notably, the ReLU activation function performed less efficiently, exhibiting higher MSE values when using five neurons compared to the other two functions. Therefore, the study concluded that a single hidden layer with five neurons provides an optimal balance between computational simplicity and prediction accuracy.

Influence of manipulated variables

The study also examined the effect of manipulated variables, such as reflux ratio, lateral outflow, and reboiler duty, on model performance. Utilizing the RMSPROP algorithm with a linear activation function, the exclusion of reboiler duty led to the poorest prediction outcomes, whereas including all three variables or excluding only the lateral outflow yielded the best results. This highlights the critical importance of reboiler duty and reflux ratio in driving the dynamics of the separation process. The lateral outflow showed relatively less influence, which can be attributed to its smaller flow rate and the reduced effect it has on the overall process.

These findings align with insights from Segovia-Hernández et al. (2004), who noted that in DWC systems, the lateral flowrate becomes less significant when the intermediate component quantity is small. Under these conditions, process dynamics are primarily governed by the reflux ratio and reboiler duty, which help mitigate disturbances affecting the lateral outlet. Furthermore, the large column diameter inherent in DWC designs contributes to enhanced process controllability and stability by minimizing the impact of fluctuations.

Error analysis and ANN performance

Error analysis using the RMSPROP algorithm, with networks ranging from 1 to 50 neurons, confirmed that a network with five neurons per layer offers optimal predictive performance with minimal error. Increasing the number of layers led to higher levels of complexity and error, emphasizing the efficiency of simpler ANN architectures in this application.

The ANN model demonstrated strong predictive capabilities for independent validation data generated by Aspen Dynamics, successfully predicting the mass fractions of acetone, ethanol, and butanol. Fig. 9.4 showcases the close alignment between ANN predictions and the simulation data, further validating the robustness of the model.

Implications for process intensification and sustainability

This study highlights the potential of using ANN modeling to optimize intensified separation processes in biobutanol production. By leveraging machine learning techniques to identify minimalistic architectures, industries can optimize their processes while minimizing energy consumption. The integration of advanced thermodynamic models with neural network optimization offers a scalable and efficient framework for solving complex separation problems in biobutanol production.

The significance of this approach extends beyond biobutanol production, as it can be applied to a variety of biofuel and sustainable chemical manufacturing processes. The reduction in computational and operational demands achieved by this methodology aligns with global sustainability goals, including those outlined in the United Nations 2030 Agenda.

In conclusion, the combined application of process intensification and ANN modeling offers a transformative pathway for sustainable chemical engineering. This approach not only enhances process efficiency but also fosters the broader adoption of renewable energy technologies, paving the way for a greener and more sustainable future. The study exemplifies the role of advanced process systems in driving innovations that contribute to achieving the sustainability targets set by the global community.

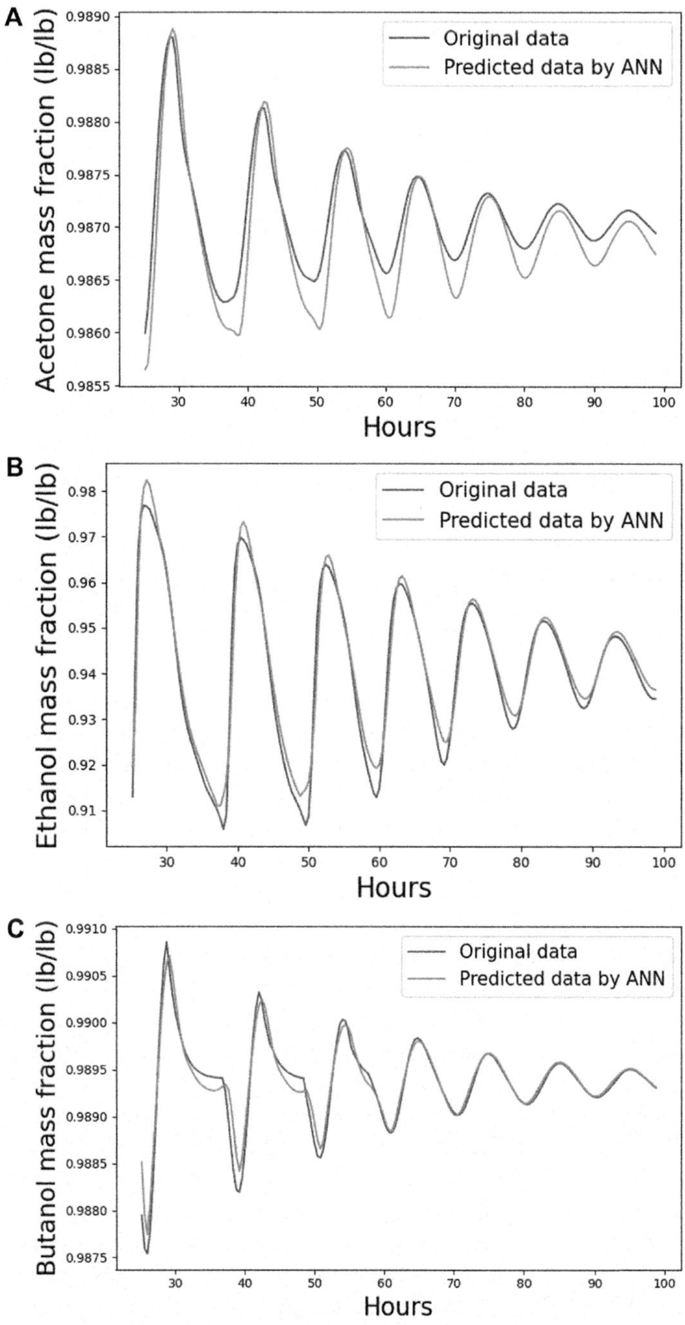

FIGURE 9.4 Prediction of the chosen ANN architecture. Comparison between the predicted and actual (a) acetone, (b) ethanol, and (c) butanol mass fraction.

Conclusions

The conclusions of this work underline the introduction of a statistical methodology aimed at identifying a minimalistic architecture for ANNs. The methodology was demonstrated using a sustainable and intensified chemical process, specifically the DWC. Through numerical analyses, it was shown that effective process prediction can be achieved by employing only two critical features: the reflux ratio and the reboiler duty. This result contrasts with previous literature, emphasizing that a single hidden layer with as few as five neurons is sufficient to model the dynamics of complex distillation columns. This simplicity is vital in avoiding overfitting, a common challenge in machine learning applications.

The significance of this methodology extends beyond mere process optimization. By narrowing down the necessary input features and ANN architecture, the approach simplifies data collection and computational efforts, making it more accessible for industrial applications. Additionally, the minimalistic model serves as a robust tool for real-time monitoring and control, particularly in scenarios where computational resources or data availability are constrained. From a broader perspective, this study exemplifies the seamless integration of machine learning techniques with advanced process intensification strategies. The DWC, as a case study, demonstrates the potential of such hybrid approaches to deliver systematic and efficient frameworks for optimizing chemical engineering processes. Beyond the immediate context, this pioneering methodology lays a foundation for enhancing the precision and efficiency of neural network design in a wide range of applications. The implications of this work resonate with global sustainability initiatives, particularly those aligned with the United Nations 2030 Agenda. By leveraging the power of machine learning in conjunction with intensified processes, industries are better positioned to optimize operations, reduce environmental footprints, and tackle global challenges such as climate change and resource scarcity. This alignment with the SDGs underscores the transformative potential of these methodologies in fostering economic growth that is both inclusive and environmentally responsible. Looking ahead, the adoption and adaptation of this methodology across various industrial sectors hold immense promise. The ability to derive actionable insights from minimal input data not only accelerates decision-making processes but also enhances operational efficiency and sustainability. By advancing the integration of machine learning and process intensification, this approach paves the way for innovative solutions that address both current and emerging challenges in chemical engineering and beyond. The work thus sets a precedent for future research and development, encouraging the exploration of similar methodologies in diverse applications, ultimately contributing to a more sustainable and equitable future for our planet.

References

Biermann, F., Kanie, N., & Kim, R. E. (2017). Global governance by goal-setting: The novel approach of the UN sustainable development goals. *Current Opinion in Environmental Sustainability, 26–27*, 26–31. https://doi.org/10.1016/J.COSUST.2017.01.010

Biyanto, T. R., Widjiantoro, B. L., Abo Jabal, A., & Budiati, T. (2010). Artificial neural network based modeling and controlling of distillation column system. *International Journal of Engineering, Science and Technology, 2*(6).

Bulsari, A. B., & Palosaafi, S. (1993). Neural computing & applications application of neural networks for system identification of an adsorption column. *Neural Computing and Applications, 1*, 160–165.

Cavanaugh, J. E., & Neath, A. A. (2019). The akaike information criterion: Background, derivation, properties, application, interpretation, and refinements. *Wiley Interdisciplinary Reviews: Computational Statistics, 11*(3). https://doi.org/10.1002/wics.1460

Errico, M., Sanchez-Ramirez, E., Quiroz-Ramìrez, J. J., Rong, B. G., & Segovia-Hernandez, J. G. (2017). Multiobjective optimal acetone-butanol-ethanol separation systems using liquid-liquid extraction-assisted divided wall columns. *Industrial and Engineering Chemistry Research, 56*(40), 11575–11583. https://doi.org/10.1021/ACS.IECR.7B03078/SUPPL_FILE/IE7B03078_SI_001.PDF

Gómez-Castro, F. I., Segovia-Hernández, J. G., Hernández, S., Gutiérrez-Antonio, C., & Briones-Ramírez, A. (2008). Dividing wall distillation columns: Optimization and control properties. *Chemical Engineering and Technology, 31*(9), 1246–1260. https://doi.org/10.1002/CEAT.200800116

Lu, Y., Nakicenovic, N., Visbeck, M., & Stevance, A. S. (2015). Policy: Five priorities for the UN sustainable development goals. *Nature, 520*(7548), 432–433. https://doi.org/10.1038/520432a

Luyben, W. L. (1992). Practical distillation control. In *IEE colloquium (digest)* (1st ed.). Van Nostrand Reinhold. https://doi.org/10.1049/ic:20000114. Issue 23.

Nar, K., & Sastry, S. S. (2019). Persistency of excitation for robustness of neural networks. *arXiv preprint arXiv:1911.01043.*

Narendra, K. S., & Annaswamy, A. M. (2007). *Persistent excitation in adaptive systems* (Vol 45). https://doi.org/10.1080/00207178708933715

Ogunmakinde, O. E., Egbelakin, T., & Sher, W. (2022). Contributions of the circular economy to the UN sustainable development goals through sustainable construction. *Resources, Conservation and Recycling, 178*, Article 106023. https://doi.org/10.1016/J.RESCONREC.2021.106023

Rewagad, R. R., & Kiss, A. A. (2012). Dynamic optimization of a dividing-wall column using model predictive control. *Chemical Engineering Science, 68*(1), 132–142. https://doi.org/10.1016/J.CES.2011.09.022

Ribeiro, N., Pinto, A. C., Quintella, C. M., da Rocha, G. O., Teixeira, L. S. G., Guarieiro, L. L. N., Rangel, M. do C., Veloso, M. C. C., Rezende, M. J. C., da Cruz, R. S., de Oliveira, A. M., Torres, E. A., & de Andrade, J. B. (2007). The role of additives for diesel and diesel blended (ethanol or biodiesel) fuels: A review. *Energy and Fuels, 21*(4), 2433–2445. https://doi.org/10.1021/EF070060R/ASSET/IMAGES/LARGE/EF070060RF00003.JPEG

Segovia-Hernández, J. G., Hernández, S., Rico-Ramírez, V., & Jiménez, A. (2004). A comparison of the feedback control behavior between thermally coupled and conventional distillation schemes. *Computers and Chemical Engineering, 28*(5), 811–819. https://doi.org/10.1016/J.COMPCHEMENG.2004.02.019

Shin, Y., Smith, R., & Hwang, S. (2020). Development of model predictive control system using an artificial neural network: A case study with a distillation column. *Journal of Cleaner Production, 277*, Article 124124. https://doi.org/10.1016/J.JCLEPRO.2020.124124

Chapter 10

Challenges, opportunities in the improvement of complex DWC schemes

Background

Dividing Wall Columns (DWCs) are recognized as one of the most significant advancements in distillation technology, offering transformative benefits in energy efficiency, cost savings, and process intensification. They integrate multiple distillation steps within a single unit, reducing the need for additional equipment and auxiliary systems. Despite their potential, DWCs face technical and operational challenges that can hinder widespread adoption.

This chapter explores these challenges and the corresponding opportunities for improvement, focusing on strategies and topological adjustments aimed at optimizing performance. Additionally, the chapter highlights trends in the application of thermal coupling (TC), including Total Thermal Coupling (TTC), and their role in producing high-value compounds. Real-world data and case studies provide insights into the successful industrial implementation of sustainable distillation systems.

Challenges and opportunities in DWC schemes

DWCs are among the most impactful process intensification technologies in modern chemical engineering. They streamline distillation by combining multiple separation tasks within a single unit, reducing equipment count and energy consumption. However, the successful implementation of DWCs presents several challenges, from design complexity to scalability. Conversely, these challenges offer opportunities for innovation in technology and practice.

Challenges in DWC schemes

Design complexity

The design of DWCs requires a sophisticated understanding of thermodynamics, fluid mechanics, and mass transfer. Unlike conventional distillation

Thermally Coupled Distillation Columns. https://doi.org/10.1016/B978-0-443-33114-5.00005-3

columns, which typically manage binary separations, DWCs can handle three or more components, necessitating advanced computational tools.

- Quantitative Insight: The modeling of a standard DWC involves solving over 100 coupled nonlinear equations to balance mass, energy, and component distribution (Weinfeld et al., 2018). This complexity is exacerbated by the need to account for multiple interdependent parameters, such as reflux ratios, feed locations, and the positioning of the dividing wall.
- Example: In a study of a DWC separating a ternary mixture of methanol, ethanol, and propanol, the simulation required over 200 hours of computation to optimize all operating parameters using commercial software tools.
- Impact:
  Design inaccuracies can result in:
  Reduced Efficiency: A 5% error in wall placement can lead to a 10% decrease in separation efficiency (Dejanović et al., 2010).
  Energy Penalties: Suboptimal configurations can increase reboiler duties by 15% compared to optimized setups.
- Solution and Opportunity:
  The integration of machine learning (ML) into simulation platforms is showing promise. ML algorithms can analyze historical data and predict optimal configurations, reducing design time by up to 50% and improving accuracy (Long & Lee, 2014).

Operational challenges

DWCs must perform consistently across varying feed compositions, thermal loads, and product specifications, which complicates their operation.

- Case Study: A refinery processing 1000 tons/day of a ternary hydrocarbon feed encountered fluctuations in feed composition of ±5%. The lack of adaptive controls resulted in energy penalties of up to 10% and product purity reductions of 3% (Buitimea-Cerón et al., 2021).
- Example of Challenge: Non-uniform liquid and vapor distribution across the dividing wall can cause maldistribution, reducing efficiency by 12%–15% in a 3-m diameter DWC.
- Solution and Opportunity:
  Real-Time Monitoring: Advanced sensor networks integrated into DWCs can detect variations in feed conditions, enabling adaptive control mechanisms.
  Control Algorithms: The implementation of model predictive control (MPC) improved operational stability by reducing disturbances by 30% in a case study involving a pharmaceutical DWC (Sidek & Othman, 2019).

Scaling up

While DWCs have demonstrated efficacy in pilot and laboratory settings, scaling up for industrial applications remains a significant challenge.

- Key Issues in Scaling Up:
  Mechanical Constraints: Larger diameters increase the risk of maldistribution, with efficiency losses of up to 15% observed in columns exceeding 2 m in diameter (Kenig, 2022).
  Thermal Challenges: Ensuring even heat distribution becomes more complex as column height increases.
  Economic Trade-Offs: Scaling up requires more robust materials, which can raise capital costs by 20%–30%.

- Solution and Opportunity:
  Modular Designs: Modular DWCs enable efficient scaling without sacrificing performance.
  Example: A modular DWC designed for bioethanol production achieved a 25% reduction in construction costs while maintaining 90% separation efficiency for ethanol from a five-component feed.

Control systems

The integrated nature of DWCs, with multiple interacting phases and streams, makes control systems particularly challenging.

- Case Study: Traditional PID controllers exhibited a 20% longer settling time compared to MPC systems when managing thermal disturbances in a DWC (Lee et al., 2016).
- Example of Advancement: AI-driven control systems reduced energy consumption by 12% in a commercial-scale DWC producing specialty chemicals.
- Solution and Opportunity:
  The integration of AI and digital twins—virtual replicas of physical systems—can predict and preempt operational inefficiencies, minimizing downtime and improving product quality.

Opportunities in DWC schemes

Energy savings

DWCs are widely recognized for their potential to reduce energy consumption.

- Case Study: A petrochemical plant in Germany replaced a three-column system with a single DWC. The change reduced energy usage by 40%, saving 20,000 MW h/year and cutting CO_2 emissions by 9000 tons annually (Ränger et al., 2025).
- Quantitative Insight: DWCs consume 30%–50% less energy than traditional distillation systems for comparable separation tasks.

Cost reduction

By combining separation processes, DWCs reduce the need for additional equipment, such as reboilers and condensers.

- Data Point: A large-scale refinery in the Middle East retrofitted its separation system with a DWC, saving \$2.5 million in equipment costs and reducing operational expenses by 30% (Ränger et al., 2019).
- Long-Term Savings: Despite a 20% higher initial investment, the payback period for DWC installations is often less than 3 years due to reduced energy and maintenance costs.

Sustainability benefits

DWCs support global sustainability goals by reducing energy consumption, greenhouse gas emissions, and resource use.

- Quantitative Insight: A DWC used in bioethanol production reduced water usage by 20% and CO_2 emissions by 15% compared to conventional columns (Kiss, 2013).
- Future Potential: Integrating DWCs with renewable energy sources, such as solar-assisted distillation, can further reduce fossil fuel dependency.

Process intensification

DWCs align with the principles of process intensification, which aim to make chemical processes more compact and efficient.

- Example: A pharmaceutical company integrated a DWC into its purification process, reducing plant footprint by 30% and achieving 98% product purity with a 25% reduction in energy consumption.
- Quantitative Insight: Compact DWCs save 20%–40% in installation space compared to conventional multi-column systems, enabling their deployment in urban and resource-constrained settings.

DWCs offer transformative potential for the chemical processing industry by delivering significant energy savings, cost reductions, and environmental benefits. However, their implementation is not without challenges. Addressing issues such as design complexity, operational flexibility, scalability, and control requires continuous innovation in simulation tools, materials, and automation technologies.

Emerging trends such as modular designs, AI-driven controls, and renewable energy integration position DWCs as a key technology for sustainable industrial practices. By overcoming these challenges and leveraging the opportunities, DWCs are set to play a pivotal role in the future of process intensification.

Trends in the use of thermal couplings and total thermal couplings

The use of TC and TTC strategies in DWCs has revolutionized the approach to distillation, particularly in industries focused on energy efficiency and sustainability. By optimizing energy integration across different sections of the column, these strategies not only reduce energy consumption but also increase process flexibility and throughput. This section explores the latest trends in TC and TTC, highlighting key advancements and applications with supporting quantitative data.

Thermal coupling strategies: Evolution and impact on DWC performance

Thermal coupling strategies involve the integration of heat exchange between different parts of the distillation column, optimizing the energy usage and reducing redundant heating or cooling duties. Traditionally, each section of a distillation column requires its own reboiler and condenser. Thermal coupling allows for heat recovery and redistribution, improving the overall efficiency of the system (Segovia-Hernández et al., 2021).

Enhanced energy integration

One of the primary benefits of TC is the reduction of energy consumption. By reusing heat from one section of the column in another, TC minimizes the need for additional heating or cooling, lowering the reboiler duty and condenser load.

- Quantitative Insight: A case study conducted in a chemical plant in Texas demonstrates the effectiveness of TC. The plant modified a conventional distillation system separating benzene, toluene, and xylene with a TC strategy in its DWC. The result was a 25% reduction in reboiler duty and a savings of $500,000 annually in energy costs.
- Impact: Energy savings of this magnitude can significantly reduce operational costs and the environmental footprint of a facility. The reduction in energy usage not only lowers operating costs but also decreases CO_2 emissions associated with fossil fuel consumption.

Production of high-value compounds

Thermal coupling in DWCs is also being leveraged for separating complex mixtures, particularly in industries that require the production of high-value compounds, such as pharmaceuticals and fine chemicals. These industries often require the separation of compounds at very high purity levels, making efficient distillation systems even more critical.

- Case Study: In the pharmaceutical industry, a DWC used to separate an ethanol-water-acetone mixture was retrofitted with TC. The new configuration improved product purity by 5% while simultaneously reducing energy costs by 30%. This improvement was achieved by reusing thermal energy more effectively between different sections of the column, ensuring more precise temperature control during the separation process.
- Quantitative Insight: The TC retrofit helped the plant reduce energy costs by $1.2 million annually while maintaining the desired product purity levels. The process efficiency was increased, resulting in fewer rejects and higher-quality yields.

Trends in total thermal coupling (TTC)

While TC optimizes energy flows between two or more sections of a distillation column, TTC takes this concept further by integrating all energy streams throughout the entire column system. TTC eliminates redundancy in heat exchange and maximizes heat recovery, leading to even greater efficiency in energy usage. As the next frontier in process intensification, TTC is seeing widespread adoption in industries aiming for zero-energy waste and maximum thermal efficiency.

Industrial adoption of TTC

TTC systems are increasingly being adopted in large-scale industrial applications, particularly in refineries and chemical plants. These systems offer substantial energy savings and operational cost reductions by ensuring that energy is used more effectively across all stages of the distillation process.

- Case Study: A large refinery in India replaced two traditional distillation columns with a TTC-enabled DWC to separate crude oil fractions. The new TTC system resulted in a 20% reduction in overall operational costs and a 35% reduction in energy consumption. The system was also more compact, occupying 30% less space than the previous setup.
- Quantitative Insight: The refinery's investment in TTC technology resulted in annual energy savings of approximately 40,000 MW h, reducing fuel consumption and cutting CO_2 emissions by 25%. Such savings contribute not only to financial gains but also to meeting increasingly stringent environmental regulations (Yu et al., 2024).

Integration with renewable energy

Another significant trend in TTC systems is their integration with renewable energy sources, such as solar thermal energy. By utilizing renewable energy for heating or cooling duties, these systems can drastically reduce reliance on fossil fuels, contributing to a more sustainable industrial process.

- Quantitative Insight: A pilot study in Spain demonstrated the integration of solar thermal energy with a TTC system in a chemical plant. The use of solar energy to assist the TTC system reduced fossil fuel dependency by 50%, allowing the plant to cut its carbon footprint by 1500 tons of CO_2 per year.
- Example: In this pilot, solar-assisted TTC was used to heat a distillation column, with the solar system meeting 60% of the column's thermal energy demand. The solar-assisted process resulted in a 25% reduction in overall energy consumption for the plant. This integration of renewable energy highlights the potential for TTC systems to contribute to global sustainability goals (Yu et al., 2024).

Scalability and flexibility in modular TTC systems

As industries face the increasing need for more flexible and scalable production units, modular TTC systems have emerged as a practical solution. These systems allow for easier installation and the ability to handle varying feedstocks without extensive downtime or costly modifications.

- Example: A bioethanol production plant installed a modular TTC-enabled DWC that achieved a 90% efficiency rate in separating ethanol from a five-component feedstock. The modular design allowed the system to be easily scaled up or down based on production demand, increasing flexibility.
- Quantitative Insight: The modular TTC system reduced the plant's energy consumption by 20% compared to conventional distillation systems, and it increased throughput by 15%. This flexibility also made the plant more resilient to fluctuations in feedstock availability, contributing to its long-term profitability.

Reduction of equipment footprint and operational costs

One of the key advantages of TTC systems is the reduction in the equipment footprint. Because energy is reused more effectively, fewer pieces of equipment are required, leading to savings in both installation and maintenance costs.

- Case Study: A chemical plant in the United States incorporated a TTC system to separate a mixed alkylate product. The retrofitted system reduced the number of columns, reboilers, and condensers required, leading to an equipment cost savings of $1.5 million.
- Quantitative Insight: The reduction in equipment also resulted in a 10% decrease in maintenance costs, as the plant required fewer components subject to wear and tear. Furthermore, the TTC system reduced annual energy costs by $600,000, resulting in a return on investment within 3 years of installation (Yildirim et al., 2011).

Advantages and future directions

The trends in TC and TTC technologies are driving the future of sustainable and efficient distillation processes. These systems enable industries to significantly reduce their energy consumption, lower operational costs, and improve sustainability metrics. By providing better energy integration, reducing reliance on fossil fuels, and increasing scalability, TTC systems are positioned to be integral to the next generation of chemical and refining processes.

Quantitative insights into future trends

Energy consumption reduction

TTC-enabled systems are expected to reduce energy consumption by an average of 30%–40% in large-scale industrial applications.

- Forecasted Savings: By 2030, the widespread adoption of TTC systems could reduce global industrial energy consumption by 2%–3%, equating to savings of over 100 million MWh annually across the chemical and refining sectors (Yildirim et al., 2011).

Operational cost savings

- Projected Reduction: Companies implementing TTC technologies could see operational cost reductions of up to 30%, primarily from energy savings, equipment consolidation, and reduced maintenance needs.
- Cost Impact: In large-scale refineries, this could result in savings of $10 million to $15 million per year, depending on the size and scope of the operation.

Environmental impact

The integration of renewable energy sources into TTC systems is expected to further decrease the carbon footprint of industrial distillation processes. By 2030, the adoption of solar-thermal assisted TTC could lower CO_2 emissions in chemical plants by up to 5 million tons annually.

The integration of TC and TTC in distillation systems is setting a new standard for energy efficiency, sustainability, and operational flexibility. From reducing energy consumption by 25% to enabling the use of renewable energy, these technologies are reshaping the industrial distillation landscape. As the adoption of TTC systems continues to grow, the future promises even greater advancements in both process efficiency and environmental sustainability. By continuing to innovate and expand the use of TTC, industries can significantly improve their economic and ecological performance, contributing to the broader goals of energy conservation and carbon footprint reduction (Yildirim et al., 2011).

Conclusions

The ongoing evolution and optimization of Dividing Wall Column (DWC) schemes represent a pivotal development in the drive toward more sustainable and energy-efficient chemical processes. DWCs, by virtue of their innovative design, allow for significant energy reductions and cost savings, particularly when compared to conventional distillation methods. Despite their proven advantages, the implementation and operation of DWCs present numerous challenges, including complexities in design, operational flexibility, and scalability. However, these challenges have spurred considerable innovation in areas such as simulation tools, control strategies, and system design, ultimately enhancing the viability of DWCs for both small-scale and large-scale industrial applications.

This section consolidates the key findings from recent advancements in DWC schemes, focusing on their role in reducing energy consumption, capital expenditures, and operational costs. Furthermore, the integration of TC and TTC has added an entirely new dimension of efficiency, offering even greater opportunities for process intensification. As we look toward the future, it is evident that DWCs, enhanced by cutting-edge technologies such as AI-driven process optimization and renewable energy integration, will continue to be at the forefront of efforts to reduce the environmental impact of industrial processes.

Challenges driving innovation in DWC schemes

The rapid development of DWC schemes is fundamentally driven by the need to address various operational challenges that have traditionally hindered distillation processes. These challenges include:

- Design Complexity: The integration of multiple separation stages in a single column requires advanced simulation tools and precise thermodynamic modeling. Achieving an optimal design is not only a matter of technical expertise but also a balance between capital expenditure and operational efficiency. A typical DWC design process involves solving hundreds of simultaneous equations that govern mass and energy balances for multiple components. A failure to accurately model these parameters can result in suboptimal separation and wasted energy. This complexity is exacerbated when scaling up from laboratory or pilot scale to full industrial scale, which often introduces mechanical constraints such as uneven liquid/vapor distribution and increased pressure drops.
- Operational Flexibility: DWCs are designed to handle varying feed compositions, product specifications, and thermal loads. As industry trends move toward more flexible production models—driven by the need to address dynamic market demands and fluctuating raw material availability—DWCs must adapt to these changing conditions. For instance, a

variation in feed composition by ±5% could lead to an energy penalty of up to 10% if the system is not properly optimized. Thus, ensuring stable operation across a wide range of conditions remains a significant operational hurdle.

- Scaling Issues: Scaling up a DWC from a pilot or laboratory scale to full industrial scale introduces challenges such as maldistribution of vapor and liquid phases, which can reduce efficiency by up to 15%. Furthermore, large-scale systems demand more advanced control systems and mechanisms to manage the interactions between multiple streams and phases.

However, these challenges have not deterred progress. In fact, they have spurred the development of innovative solutions that enhance the design, control, and operational efficiency of DWCs. With the advent of advanced simulation tools, real-time monitoring systems, and adaptive control algorithms, the once daunting task of designing and scaling DWCs has become more achievable.

Technological advancements in thermal coupling and TTC

One of the most significant advancements in DWC schemes has been the integration of TC and the more advanced TTC. These technologies offer profound improvements in energy efficiency by optimizing heat flows across various sections of the column. Rather than relying on traditional heat exchangers, which require separate heating and cooling units for each section, TC strategies allow for the recycling of heat within the system, reducing the need for external heating or cooling.

Thermal coupling

The concept of TC involves directly transferring thermal energy between different sections of the column, facilitating more efficient heat distribution. In this context, a key advantage is the reduction in the reboiler duty, which directly translates into energy savings. A notable case study in a chemical plant in Texas demonstrated that incorporating TC in a DWC reduced reboiler duty by 25%, resulting in an annual energy cost saving of $500,000 (Waibel et al., 2024).

In addition to energy savings, TC in DWCs also enhances process control by enabling better temperature management. The ability to control temperatures more precisely helps meet strict product purity requirements, especially in industries like pharmaceuticals and petrochemicals, where product specifications are often stringent.

Total thermal coupling (TTC)

TTC represents the next level of energy integration. By completely integrating the thermal energy exchange between all sections of the distillation column,

TTC eliminates redundancies in heat management, thus maximizing energy recovery and significantly reducing energy consumption. The adoption of TTC has proven especially beneficial in large-scale applications where energy costs are a major operational concern.

A case study in India's petrochemical industry highlighted the benefits of TTC when a refinery replaced two traditional distillation columns with a TTC-enabled DWC. This upgrade led to a 35% reduction in energy consumption and a 20% decrease in operational costs. The integration of TTC not only reduced the need for external energy inputs but also decreased the equipment footprint, reducing both capital and maintenance costs.

Moreover, the integration of renewable energy sources such as solar thermal systems into TTC schemes is becoming more common. A pilot study conducted in Spain demonstrated that a solar-thermal-assisted TTC system reduced the plant's fossil fuel dependency by 50%, significantly decreasing both energy costs and CO_2 emissions. The renewable energy integration into TTC systems allows for a more sustainable, low-carbon distillation process, which is becoming increasingly important in the context of global environmental goals (Segovia-Hernández et al., 2021).

Key benefits and quantitative insights

The integration of DWC, TC, and TTC offers substantial benefits across a range of operational and environmental metrics.

Energy savings

DWCs have the potential to reduce energy consumption by as much as 50% compared to conventional distillation configurations. In particular, the energy savings come from the elimination of multiple columns, reboilers, and condensers, as well as the integration of heat recovery systems.

- Example: Retrofitting a refinery with a DWC and TC system can reduce annual energy consumption by up to 40,000 MW h, leading to a substantial decrease in fuel costs and CO_2 emissions (Asprion & Kaibel, 2010).

Cost reduction

Capital Expenditure: Retrofitting an existing distillation system with a DWC and TC can save millions in capital expenditures. A typical retrofit project may result in savings of up to $5 million in equipment costs alone, thanks to the reduction in the number of columns, reboilers, and heat exchangers.

- Operational Costs: Energy cost reductions are the most obvious financial benefit of DWCs. A case study in the petrochemical industry revealed a 30% reduction in operational costs, equating to annual savings of $2.5 million.

Environmental sustainability

Reduction in CO_2 Emissions: DWC and TTC systems significantly reduce carbon emissions by improving energy efficiency and integrating renewable energy sources. For instance, a refinery using solar-thermal-assisted TTC was able to cut its CO_2 emissions by 25%, which translates to a reduction of approximately 5000 tons of CO_2 per year. These improvements contribute to a more sustainable and environmentally friendly industrial process.

Scalability and flexibility

The flexibility of modular DWC systems, particularly when equipped with TTC, is a growing trend. These systems can be scaled up or down based on production requirements, making them highly adaptable for industries with fluctuating production volumes. A modular bioethanol production plant, for example, achieved an efficiency rate of 90% in separating ethanol from a five-component feedstock, with an energy consumption reduction of 20% compared to conventional distillation systems.

The future of DWC schemes

Looking ahead, the future of DWC schemes is promising. The continued development of advanced simulation tools, real-time monitoring systems, and adaptive control strategies will further enhance the design and operation of DWCs. These advancements will help overcome existing challenges, such as variability in feed composition and thermal loads, allowing for even greater energy efficiency and operational flexibility.

The integration of AI-driven process optimization will also play a crucial role. AI algorithms can predict and adjust operational parameters in real-time, optimizing energy use and minimizing process disruptions. Additionally, the increasing adoption of renewable energy sources, such as solar and geothermal, will make DWC systems even more sustainable, helping industries meet global carbon reduction targets.

Ultimately, as industrial sectors worldwide strive to meet sustainability goals, DWC schemes—enhanced by TC and TTC—will become essential in the quest for more efficient, cost-effective, and environmentally responsible chemical processing. By addressing existing challenges and capitalizing on emerging trends, DWCs are poised to transform the future of industrial distillation and process design, making them indispensable in the global effort toward a more sustainable future.

References

Asprion, N., & Kaibel, G. (2010). Dividing wall columns: Fundamentals and recent advances. *Chemical Engineering and Processing: Process Intensification, 49*(2), 139–146.

Buitimea-Cerón, G. A., Hahn, J., Medina-Herrera, N., Jiménez-Gutiérrez, A., Loredo-Medrano, J. A., & Tututi-Avila, S. (2021). Dividing-wall column design: Analysis of methodologies tailored to process simulators. *Processes, 9*(07), 1189.

Dejanović, I., Matijašević, L., & Olujić, Ž. (2010). Dividing wall column—A breakthrough towards sustainable distilling. *Chemical Engineering and Processing: Process Intensification, 49*(6), 559–580.

Kenig, E. Y. (2022). State-of-the-art modeling of separation columns: A review. *Chemical Engineering Transactions, 94*, 325–330.

Kiss, A. A. (2013). Novel applications of dividing-wall column technology to biofuel production processes. *Journal of Chemical Technology and Biotechnology, 88*(8), 1387–1404.

Lee, M., Minh, L. Q., Long, N. V. D., & Shin, J. (2016). Retrofit of side stream columns to dividing wall columns, with case studies of industrial applications. *Chemical Process Retrofitting and Revamping: Techniques and Applications*, 249–283.

Long, N. V. D., & Lee, M. (2014). Review of retrofitting distillation columns using thermally coupled distillation sequences and dividing wall columns to improve energy efficiency. *Journal of Chemical Engineering of Japan, 47*(2), 87–108.

Ränger, L. M., Halvorsen, I. J., Grützner, T., & Skogestad, S. (2025). What can go wrong in a dividing wall column and how to detect it. *Separation and Purification Technology, 354*, Article 129151.

Ränger, L. M., Preißinger, U., & Grützner, T. (2019). Multiple dividing-wall columns–current status and future prospects. *Chemie Ingenieur Technik, 91*(4), 420–428.

Segovia-Hernández, J. G., Hernández, S., Sánchez-Ramírez, E., & Mendoza-Pedroza, J. (2021). A short review of dividing wall distillation column as an application of process intensification: Geographical development and the pioneering contribution of Prof. Arturo Jimenez in Latin America. *Chemical Engineering and Processing-Process Intensification, 160*, Article 108275.

Sidek, N. M., & Othman, M. R. (2019). A review of heat integration system in dividing wall column for oleochemicals separation. *Materials Today: Proceedings, 19*, 1225–1231.

Waibel, Y., Ränger, L. M., Fischer, M., & Grützner, T. (2024). The start-up of a multiple dividing wall column–A theoretical and experimental study. *Chemical Engineering and Processing-Process Intensification, 201*, Article 109798.

Weinfeld, J. A., Owens, S. A., & Eldridge, R. B. (2018). Reactive dividing wall columns: A comprehensive review. *Chemical Engineering and Processing-Process Intensification, 123*, 20–33.

Yildirim, Ö., Kiss, A. A., & Kenig, E. Y. (2011). Dividing wall columns in chemical process industry: A review on current activities. *Separation and Purification Technology, 80*(3), 403–417.

Yu, A., Ye, Q., Li, J., Li, X., Wang, Y., & Rui, Q. (2024). Economic, environmental, energy, exergy (4E) analysis and simulated annealing algorithm optimization of dividing-wall column-intensified heterogeneous azeotropic pressure-swing distillation process. *Energy, 296*, Article 131099.

Index

Note: Page numbers followed by '*f*' indicate figures, '*t*' indicate tables.

H

I

K

L

M

Printed and bound by CPI Group (UK) Ltd, Croydon, CR0 4YY
01/09/2025
01947335-0001